CONTENTS

THE STATE OF BIODIVERSITY IN LAKE TANGANYIKA

A LITERATURE REVIEW

EDITED BY G. PATTERSON AND J. MAKIN

Contributions by (in alphabetical order): E. Allison, T. Bailey-Watts, J. Bennett, C. Cocquyt, P. Coveliers, L. De Vos, I. Downey, R. Duck, C. Foxall, K. Goudeswaard, M. Holland, K. Irvine, K. Martens, J. McManus, G. Patterson, I. Payne, L. Risch, J. Snoeks and N. Wiltshire

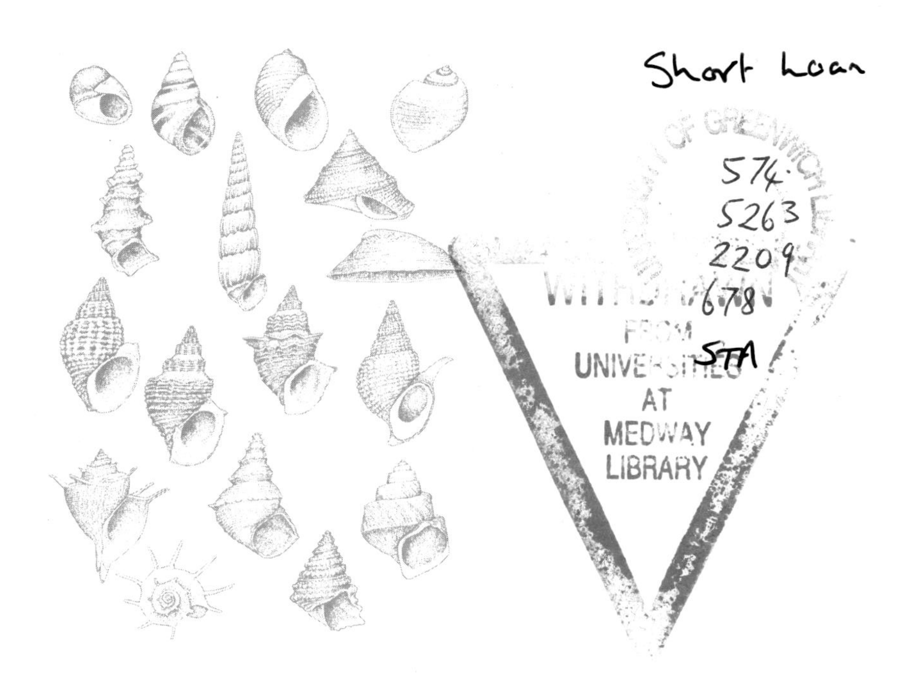

BURUNDI D.R. CONGO TANZANIA ZAMBIA

The Natural Resources Institute (NRI) is a scientific institute within the University of Greenwich, and is an internationally recognized centre of expertise in research and consultancy in the environment and natural resources sector. Its principal aim is to increase the productivity of renewable natural resources in developing countries in a sustainable way by promoting development through science.

The State of Biodiversity in Lake Tanganyika – A Literature Review was published for the Lake Tanganyika Biodiversity Project by the Natural Resources Institute

Cover photograph	Much of the subsistence fishing effort on the lake is carried out close inshore from small canoes
Title page illustration	Some gastropods of Lake Tanganyika showing the variety of forms (D. Voorwelt)

Copies of this book can be obtained by writing to NRI Catalogue Services, CAB International, WALLINGFORD, Oxon OX10 8DE, UK. When ordering, please quote **F10**.

Natural Resources Institute

ISBN: 0 85954 492-3

Plate 1

Plate 2

Plate 3

Plate 1. A typical village on Lake Tanganyika with nets drying on the shoreline.

Plate 2. The main catch on Lake Tanganyika is of small pelagic clupeids (sardines): these fish are generally sun dried and are traded throughout the region.

Plate 3. Deforestation on the steep slopes adjacent to the lake has resulted in an increased amount of soil erosion and therefore sediment entering the lake.

INTRODUCTION

The aim of the Lake Tanganyika Biodiversity Project is to help the riparian states produce an effective and sustainable system for managing and conserving the biodiversity of Lake Tanganyika into the foreseeable future.

Among the principal objectives of the project is the establishment of a sustainable regional management plan for pollution control, conservation and maintenance of biodiversity in Lake Tanganyika. This will be founded upon the results of a series of multi-disciplinary studies aimed at improving understanding of the complex scientific, technical, legal and socio-economic issues related to conservation of the lake and its immediate environment.

There are five nominal study components to the project. These are:

- biodiversity, to find out exactly what species and combinations of species and habitats are under particular threat;
- pollution, to identify the sources, evaluate the consequences and find preventative measures;
- sedimentation, to monitor the movement and impact of soil entering the lake;
- socio-economics and
- environmental education, two interlinked programmes intended to raise awareness of critical environmental issues among user groups, and facilitate translating the scientific studies into locally acceptable practices and policies in which local people are able to play a much greater part in conservation and development.

Incorporated under these headings are studies on fishing and agricultural practices, merits of sites for underwater national parks, the relevance of the legal systems of land ownership, lake conservation and developmental needs considering all the problems associated with the huge distances and poor communications involved.

As a first stage of this project, a team of national experts and/or consultants was appointed to compile and review the existing information on all aspects of the project. This task was completed in January 1996 and included five separate baseline reviews which were presented to the project steering committee and discussed at the project inception workshop which was held in March 1996. In the light of these discussions, some minor changes were made.

These five baseline reviews were:

- Biodiversity
- Pollution and its Effects on Biodiversity
- Sediment Discharge and its Consequences
- Social, Economic and Sectoral Features
- Legal and Institutional.

They included literature reviews of the individual topics as well as preliminary plans of the special studies based on identified gaps in knowledge, and an assessment of the physical and human resources available to the project in the region.

This publication is a compilation of the literature review section of the three technically oriented baseline reviews – Biodiversity, Sediment Discharge and its Consequences and Pollution and its Effects on Biodiversity. It was decided to compile and publish these three sections separately as together they form an excellent recent literature review of the current state of Lake Tanganyika. Wider distribution of these literature reviews has therefore been considered desirable and meets one of the principal aims of the project – to provide resource material to the institutions who have a research interest in Lake Tanganyika and its catchment.

G. Patterson
Natural Resources Institute

1. BIODIVERSITY OF LAKE TANGANYIKA

1.1 INTRODUCTION

Lake Tanganyika is the second deepest lake in the world and is most probably of the order of 10 million years old. More than 1200 species of organisms have been found so far in the lake, which gives Lake Tanganyika the second highest recorded diversity for any lake on earth (Cohen *et al.*, 1993). This alone establishes the significance of Lake Tanganyika as a centre of biodiversity. However, only about 10% of the coast line of the lake has been examined scientifically (Lowe-McConnell, 1987) and consequently the recorded biological diversity will almost certainly be increased with further studies. There can be no question that Lake Tanganyika is a site of global significance, largely due to the diversity of organisms within it.

Amongst the major groups contributing to the high levels of diversity are the fish, which have probably received the most attention. They also display a further aspect of the degree of biodiversity in the lake: the high degree of differentiation.

The most prolific group of fish belongs to the family Cichlidae and, of the 172 recorded species, 97% are endemic, that is, they have evolved in the lake. The speciation of cichlids appears to be a general feature of the Great Lakes of East Africa. A high degree of differentiation is also true of the non-cichlid species, in that of the 118 recorded species some 46% are endemic (Coulter, 1991a). Whilst less diverse than the cichlids, many non-cichlid species have a rather wider distribution than the frequently more localized cichlid types. A further feature of the differentiation of species is that evolutionary divergence has extended beyond the species and generic levels as far as families which have originated in the lake.

Such diversification is known from other groups within the lake, but it is not necessarily the case for all groups in any centre of biodiversity. Within tropical lakes as a whole, the fish most commonly produce a high diversity of species, but within phytoplankton and zooplankton assemblages this is not the case. Species number is often no higher than for temperate lakes, and there is often a high proportion of cosmopolitan and pan-tropical species (Payne, 1986). Even in a renowned centre of biodiversity such as Lake Tanganyika, therefore, groups need to be reviewed independently to establish which ones respond to environmental conditions by evolving new species.

The key to maintaining and even possibly initiating biodiversity is the nature of the habitats available within the ecosystem. Those organisms which occupy more homogeneous environments or more physiologically extreme environments tend to have a lower diversity than those which occupy habitats of greater physical complexity. This would include invertebrates amongst vegetation, algae on rocks and above all, fishes, whose mobility in three dimensions helps them to create their own complexity of environments. The organisms which experience long-lasting variety or patchiness in their environment are provided with the opportunity to express potential for a high diversity (Payne, 1986). This concept of habitat diversity has been termed the 'environmental mosaic' on the principle that the more 'pieces' there are, the more species can be supported by the habitat (Hutchinson, 1959). The types and varieties of habitats available are, therefore, significant in a review of biodiversity. In Lake Tanganyika, as in any lake, there are basically two kinds of places where animals and plants can live: on the bed of the lake, or in the overlying water. These broad areas are influenced by depth, and in the case of the lake bed by the physical nature of the bottom materials, to create a number of more-or-less well defined habitat types which in turn tend to be associated with characteristic assemblages of species. As the physical nature of the habitats is essential for the maintenance and probably also the initiation of biodiversity, a review of the more significant habitats is required, particularly because of their essential role in conservation and the future need for representative sampling.

1.2 LAKE HABITATS

1.2.1 General categories

Within Lake Tanganyika, organisms can live either on the bed of the lake or in the overlying water, but within these two broad divisions a number of more specific categories of habitat can be identified. In such a deep lake three-dimensional differentiation of habitats will clearly be a major feature in enhancing the environmental mosaic.

A number of major habitats or zones associated with depth have been identified, based on physical and biological criteria (Coulter, 1991a). These are summarized in Table 1.1. These habitats are also related to the profile of the lake to some extent, and vary both in the nature of their species assemblages and in the number of species they tend to support (Figure 1.1). The highest number of species for most groups of organisms tends to be found in the littoral rim of the lake. However, whilst depth is obviously important in defining habitats, it is not quite so influential as might be suspected from purely bathymetric data and the immense depth of the lake. At depths greater than 100–200 m, depending upon the locality, the water becomes completely anoxic (see discussion on the pelagic zone below). This means that some 75% of the water volume of the lake is uninhabitable for most types of organisms. The vertical differentiation of habitats is therefore relatively truncated by this phenomenon, given that the lake is 1460 m deep.

Table 1.1 Major habitats associated with depth

Habitat	**Depth** (m)
Open water	
Pelagic	0–40
Bathypelagic	>40
Lake bed	
Littoral	0–10
Sub-littoral	10–40
Benthic	>40

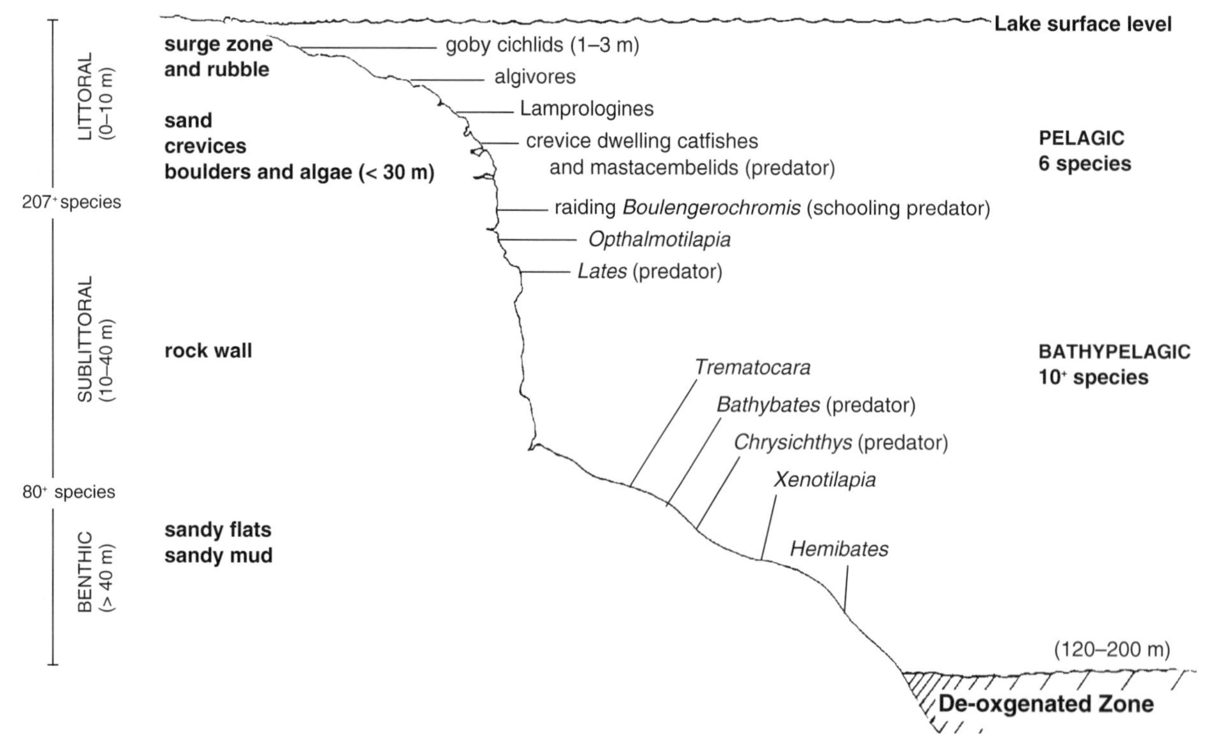

Figure 1.1 Lake Tanganyika inshore biotopes and typical fishes. (From sketch by R. H. Lowe-McConnell, November 1995.)

In addition to the intrinsic habitats of the lake, there are also those offered by the inflowing rivers, which are often associated with alluvial deltas and, in the case of the Malagarasi and Ruzizi, also with extensive swamp vegetation. In the context of the basin as a whole, these rivers make their own distinctive contribution to the biodiversity.

1.2.2 Open water

Pelagic habitats

The structure of the pelagic zone as a habitat is largely determined by the thermal characteristics of the lake and those factors which influence this. Lake Tanganyika is permanently stratified, with the depth of the thermocline varying annually and seasonally but generally to be found at depths of 100–200 m.

Unlike many tropical lakes there are rarely any secondary thermoclines found above the main one, although daily stratification of the first 5–10 m under the immediate influence of the sun may occur (Hecky, 1991). This most superficial layer, including that at the lake margins, may vary by up to 3 °C during a day (Caport, 1952; Coulter, 1968). Below this, temperatures are generally stable on a daily basis, but may show small seasonal fluctuations. Within the epilimnion proper, down to around 60 m, temperatures are generally isothermal but range from 27 °C in April during the strongest stratification to 25 °C in August (Lindquist and Mikkola, 1989). In the hypolimnion (alternatively defined as the monimolimnion by Hecky *et al.*, 1991), the temperature lies between 23.25 and 23.48 °C at various locations (Lindquist and Mikkola, 1989).

The critical feature of a thermocline is that it acts as a physico-chemical barrier to diffusion. Consequently, decomposition of the rain of debris from the surface below the thermocline liberates inorganic nutrients which then become trapped within the hypolimnion. Most significantly, however, the oxidative processes of decomposition strip the oxygen from the water of the hypolimnion, and the thermocline barrier prevents replacement by diffusion from the more superficial layers above the thermocline. This is the origin of the large volume of anoxic water below the thermocline in Lake Tanganyika, and it is this which effectively provides the lower boundary for life in both pelagic and bottom habitats. It is the second largest body of anoxic water of any inland water body after the Black Sea, and occupies around 75% of the total volume of the lake. Whilst this phenomenon is known from other tropical lakes, the relatively narrow band of the oxygenated pelagic zone is a particular feature of Lake Tanganyika (Hecky, 1991). The depth of this oxygenated zone varies along the lake. At the north end the anoxic water is quite close to the surface and occurs below 80 m. There is a gradient from north to south with the oxygenated layer penetrating progressively more deeply until, in the south-eastern basin, it reaches 200 m before anoxic conditions are encountered (Coulter, 1991a). The anoxic conditions may provide a habitat only for certain anaerobic micro-organisms, although in other lakes fishes are known which have become adapted for making temporary forays into the de-oxygenated zone (Trewavas *et al.*, 1972). Such specialized forms have yet to be confirmed in Lake Tanganyika, although there is indirect evidence that some species may do this (Coulter, 1991a).

Whilst Lake Tanganyika is permanently stratified in a way that has profound ecological significance, some mixing does take place. During the dry season and early monsoons the wind blows along the lake predominantly from the southwest. The effect of this is to drive the surface waters towards the north, thereby depressing the thermocline at the north end and generally tilting the whole thermocline. The upwelling of this nutrient-rich water from below the thermocline causes a bloom in planktonic organisms, which is ultimately passed on to the fishery in a similar fashion to that seen in Lake Malawi (Fryer and Isles, 1972). It is this seasonal flush of nutrients which is the main regenerative event in maintaining the productivity in the pelagic system. The same event is also probably responsible for a degree of mixing of oxygenated with de-oxygenated waters, to the extent that the oxygenated layer is rather deeper at the southern end of the lake, where it extends below the thermocline itself (Hecky, 1991). Once the winds drop as the rains arrive, the thermocline returns to its original position, although in doing so it may show some oscillation before settling down. Internal boundary waves or seiches may be set up at the interface of the epilimnion and thermocline. On occasion, oxygen has been detected as deep as 240 m, although this may have been in the troughs of one of these seiches (Coulter, 1991a).

One other factor which varies with depth is light. This is obviously important as light provides the energy to fuel the whole system. Light is attenuated down through the water column and ultimately there comes a depth, the compensation depth, where light energy is no longer adequate for net production to occur from the photosynthesis of the phytoplankton. These more superficial layers of water, where light is sufficient to allow net production from the primary producers, form the euphotic zone which tends to have equivalent to a light intensity around 1% of the incident energy. The actual depth will depend upon water transparency and upon the density of suspended material, including the phytoplankton. In Lake Tanganyika the depth of the mixed layer, where useful production can take place, varies from 20–60 m in the wet season and 100–150 m in the dry season, when deep mixing can take place (Hecky, 1991). The average depth of the euphotic zone for all seasons is 29 m. The depth of the euphotic zone not only defines the depth at which the phytoplankton of the pelagic zone can be productive, but also the depth of light penetration to bottom communities, where algae on rocks at these depths, for example, can also contribute to the productivity of the system as a whole.

In general, the pelagic region is constrained by thermal stratification and its dependent variable, oxygen availability. The pelagic zone appears relatively homogeneous, with little clear differentiation into separate habitats. With regard to the fishes at least, this zone is characterized by relatively few species (Figure 1.1), but makes the major contribution to the production of the system as a whole.

Bathypelagic habitat

The bathypelagic habitat is essentially the lower level of the pelagic region. It is a region where light is dim or absent, conditions are relatively stable and there is a constant rain of debris from the upper productive layers. It is characterized by the organisms found living there; present knowledge indicates that fish dominate. These fish live off the bottom and tend to have large eyes, well developed lateral line systems (presumably to compensate for poor light availability) and they tend to be feeble swimmers. They apparently feed mainly on zooplankton (Coulter, 1991a). The bottom habitat and its community are, however, poorly known. Only around 10 fish species are thought to occupy this habitat (Figure 1.1), occurring in deeper waters between 60 and 200 m.

1.2.3 Lake bed and coastline

Depth, profile and bottom type

The fundamental divisions of habitat by depth were given in section 1.2.1. The littoral zone is at the lake margins and might be considered to extend down to the first 10 m. It is most often rocky, although this may alternate with pebble or sandy beaches (Coulter, 1991a). In very few places there may be submerged or emergent vegetation. In many places, however, the lake bed is steep and rocky and plunges downwards, so that it may reach a depth of 1000 m within the same distance from the shore. These steep, rocky surfaces provide spatially diverse habitats for a large number of fish species and tend to be covered in a film of epilithic algae down to the depth of the limit of the euphotic zone. Where the bottom is not precipitous, there is often a transition from the rocky littoral region to a more even bottom or shelf of sand and mollusc shells within the sub-littoral zone. On the southern shelf this may be extended as a shell zone down to about 60 m within the benthic zone. Below this, the benthic zone is characterized by sand and pebbles to about 100 m, after which muddy sediments might appear. When the muddy sediments descend into the anoxic or deoxygenated depths they tend to become black. This tends to occur at about 200 m depth in the south and 100 m or less in the north of the lake (Coulter, 1991a).

As with the pelagic zone, oxygen concentration clearly defines the lower limit of the benthic zone. The lower limit averages around 80 m in the north and 240 m in the south, but at certain times of year, particularly during the mixing events of May–August, sporadic upwelling of anoxic water or oscillations of the oxycline can cause rapid changes. Vertical oscillations by as much as 40 m are known at this time (Coulter, 1991a). Even above the totally anoxic depths, oxygen concentrations in the lower regions of the benthic region may be quite low. In the southern end, for example, deep gill net catches took five species of fish at 212–215 m depth, where the oxygen concentration was only

1.44–0.59 g O_2/m^3 (Coulter, 1967). There is generally a marked reduction in species number below 150 m, with perhaps around 10 species featuring at these lower depths. These species must be habituated to low oxygen concentrations of 1–3 g O_2/m^3. Some may even be able to survive temporarily in anoxic water. The benthic habitat must therefore present an unstable oxygen climate, particularly in its lower depths.

The type of substratum is a major determinant of the species assemblages found in the littoral and benthic regions of the lake. In Lake Malawi (Lewis *et al.*, 1986) littoral habitats, each of which could be characterized by a different species assemblage, have been divided into:

- rocky zone
- sandy zone
- rock/sand interface
- weedy zone
- intermediate zone – gradation of rock to sand, often with pockets of weed
- reed beds.

In Lake Tanganyika the weed and reed zones are particularly infrequent. At greater depths, major substrate areas have been described as rocky, sandy or muddy (Cohen, 1992). However, organic sediments or mud appear to be sparse above 100 m.

It is important that the major habitats and their dependent characteristic communities be identified, as without this characterization appropriate conservation measures taking into account all species cannot be enacted. It is also important that the extent of major habitats should be mapped to assess their true extent and vulnerability. Such mapping may ultimately be needed on a very detailed scale in the conservation areas themselves, in order to provide nature trails and viewing opportunities for tourism and educational purposes (Figure 1.2).

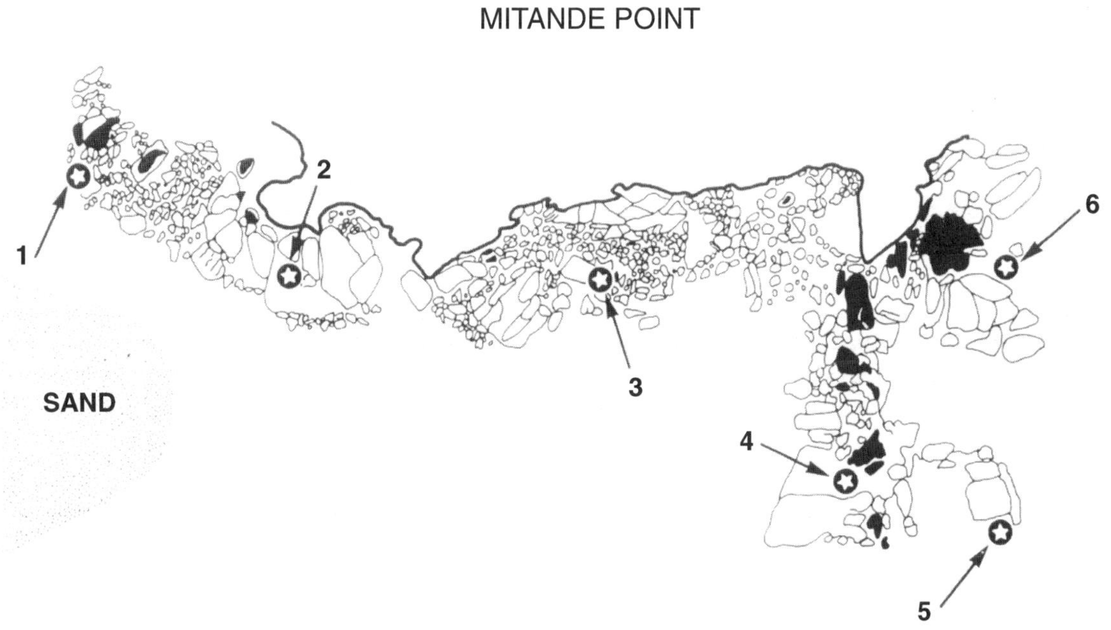

Figure 1.2 Example of underwater nature trail based on detailed habitat mapping at Cape Maclear in the Lake Malawi National Park. (From Lewis *et al.*, 1986 in conjunction with World Wildlife Fund.)

Habitat distribution basin-wide

The structure of the lake clearly has a considerable influence upon the nature of species to be found over the bottom regions of the lake. The relative distribution of habitats over the lake as a whole will therefore influence the commonness and distribution of their dependent assemblages. The type of bottom substrate will be related to the nature of the coastline, particularly the littoral zone, as well as to the steepness of the bottom.

Over 70% of the bottom is more than 200 m deep and generally unavailable for colonization due to lack of oxygen. There is no detailed habitat map of even the coastal regions of Lake Tanganyika. The total length of coastline is around 2000 km. In general the steepest areas are rocky (see Figure 2.2), and these rocky habitats are particularly associated with the western margins of the lake (Figure 1.3) and the Mahele mountains of the east.

An underwater ridge termed the Mahele shoals extends from the Mahele mountains more-or-less across the lake, effectively forming the boundary between the north and south basins of the lake. The Mahele sector is therefore at a critical point in the lake as it appears to be a good example of rocky habitat as well as providing sites in both basins. Coasts in less steep areas tend to be a mixture of rocky areas alternating with sandy bays.

There are some extensive shallower shelf areas in the lake. The two southern areas, for example, constitute around 1000 km^2 of oxygenated bottom. Most of the benthic areas here slope gently (Figures 1.4 and 2.2) and rocky areas are generally limited in extent. Other shallow areas are typically associated with sediment fans from inflowing rivers such as the Malagarasi (Figure 1.3). The inflowing rivers tend to create their own distinctive habitats. The Malagarasi, for example, enters the lake through an extensive swampy delta dominated by papyrus (*Cyperus papyrus*) and reeds *(Typha* sp.) of a type common in Lake Victoria but otherwise rare in Lake Tanganyika. The Malagarasi delta also possesses areas of submerged plants and floating vegetation which are well oxygenated and provide rich habitats for fish and invertebrates (Beadle, 1981). On a much smaller scale, similar conditions are found in the estuaries of the Ifume and other small rivers. The fine organic mud of these submerged deltas spreads out for some distance into the lake (Figure 1.3).

Vegetated areas are generally rare in Lake Tanganyika, although they appear to be very important for the early stages of juvenile fish, particularly of the species *Lates*.

Some coastline habitats are already included in or are adjacent to terrestrial conservation areas. The Ruzizi delta, for example, is incorporated into a national park which is essentially a bird reserve (Figure 1.3). The Gombe Stream National Park protects an isolated patch of forest with its chimpanzees, but the status of the rocky/sandy littoral region is unclear. There is certainly artisanal fishing activity in the 0.5-km band adjacent to the shore (K. E.Banister, personal communication).

The Mahale Mountains National Park is also rich in terrestrial vertebrate fauna including chimpanzees. The park extends over 15 km of rocky coastline and incorporates a fully protected strip of the lake. The Malagarasi river swamp and the Ugalla river swamp in the same system are incorporated into a game reserve, but not as a National Park (World Bank/DANIDA, 1994).

Nsumbu National Park in Zambia is fully designated and protection extends to the coastal regions of this shallow rocky/sandy region, where fish biodiversity is regarded as being amongst the highest in the lake (G. W. Coulter, personal communication).

Finally, it should be borne in mind that the lake occupies only a relatively small portion of the basin. The total catchment is 239 000 km^2 of which the area of the lake is only 32 000 km^2 (Figure 2.2). The Malagarasi sub-basin at 130 000 km^2 is the most extensive of the river systems, and run-off from the Tanzania sector as a whole amounts to some 60% of the total input into the lake (World Bank/DANIDA, 1994). The inflowing rivers tend to have their own flora and fauna which may be quite distinctive in their own right. Within the Malagarasi the degree of endemism may be low, but many species have affinities with the Zaire basin into which the western-flowing Malagarasi flowed before the formation of the lake (Lowe-McConnell, 1987; Banister and Clarke, 1980).

In many ways it is the riverine communities which bear the brunt of the catchment area degradation. It has been estimated from satellite imagery, for example, that up to 46% of the catchment area within the extensive Tanzanian sector has been deforested, and it is anticipated that this will directly increase

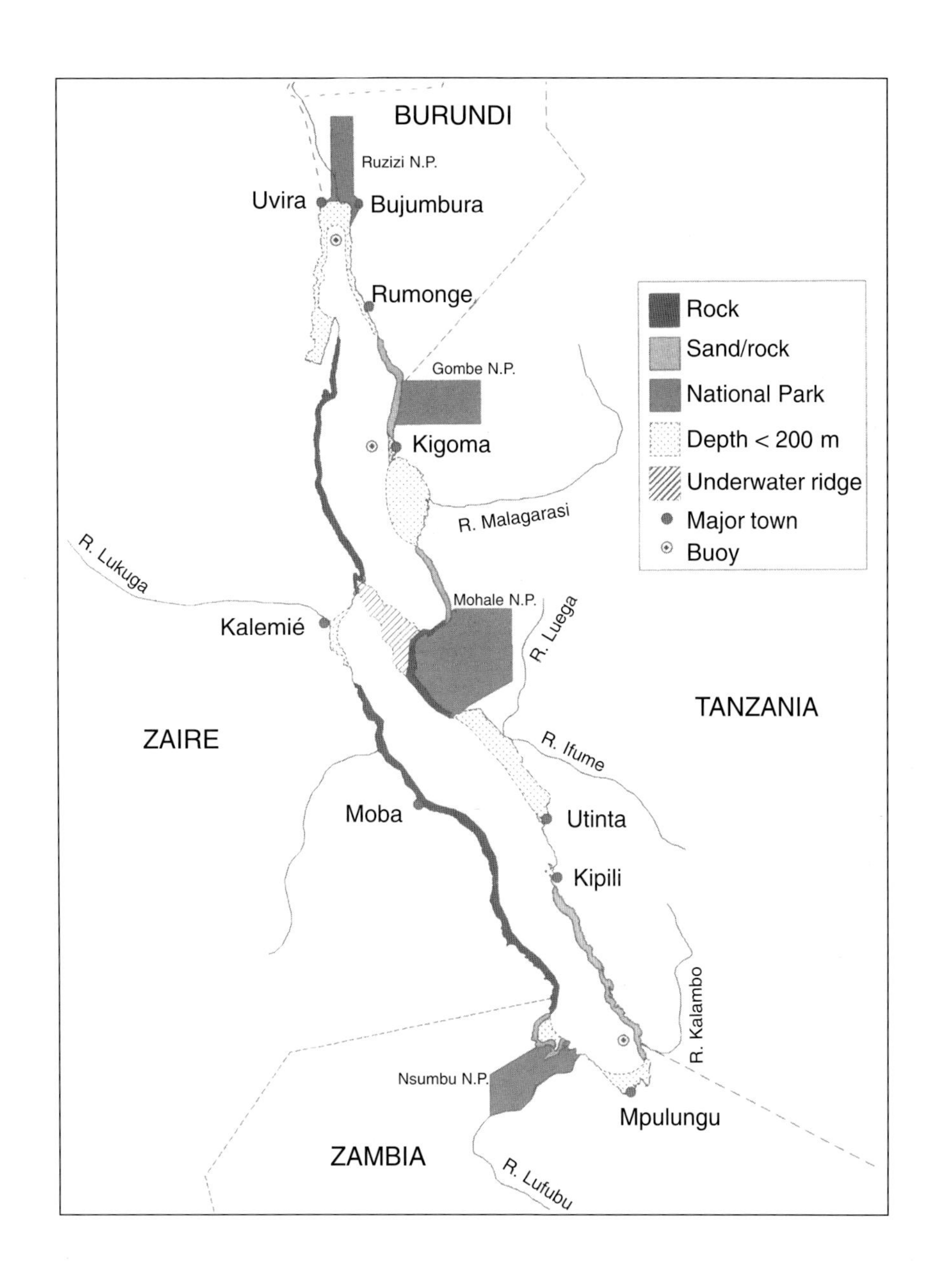

Figure 1.3 Distribution of major coastal habitat types. (From sketch by G.Coulter and R.H. Lowe-McConnell, November 1995.)

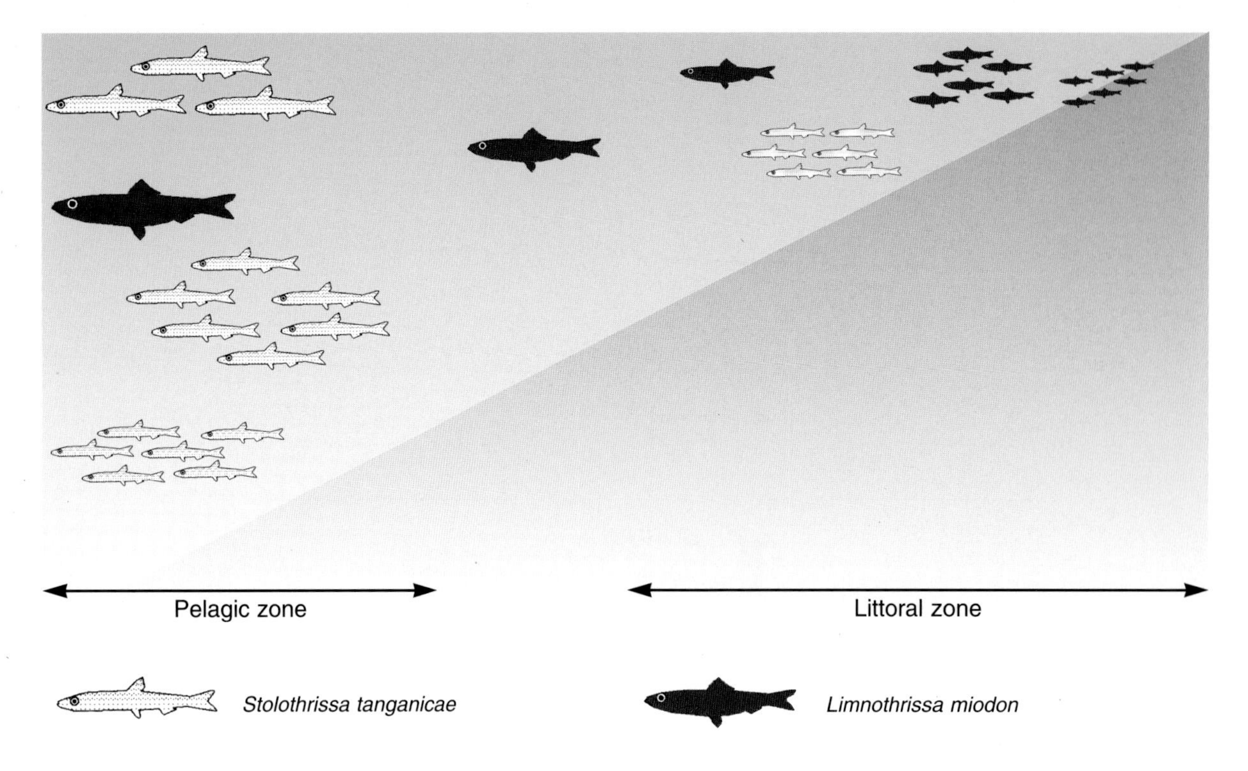

Figure 1.4 Distribution of clupeids with distance from the shore (Petit, unpublished)

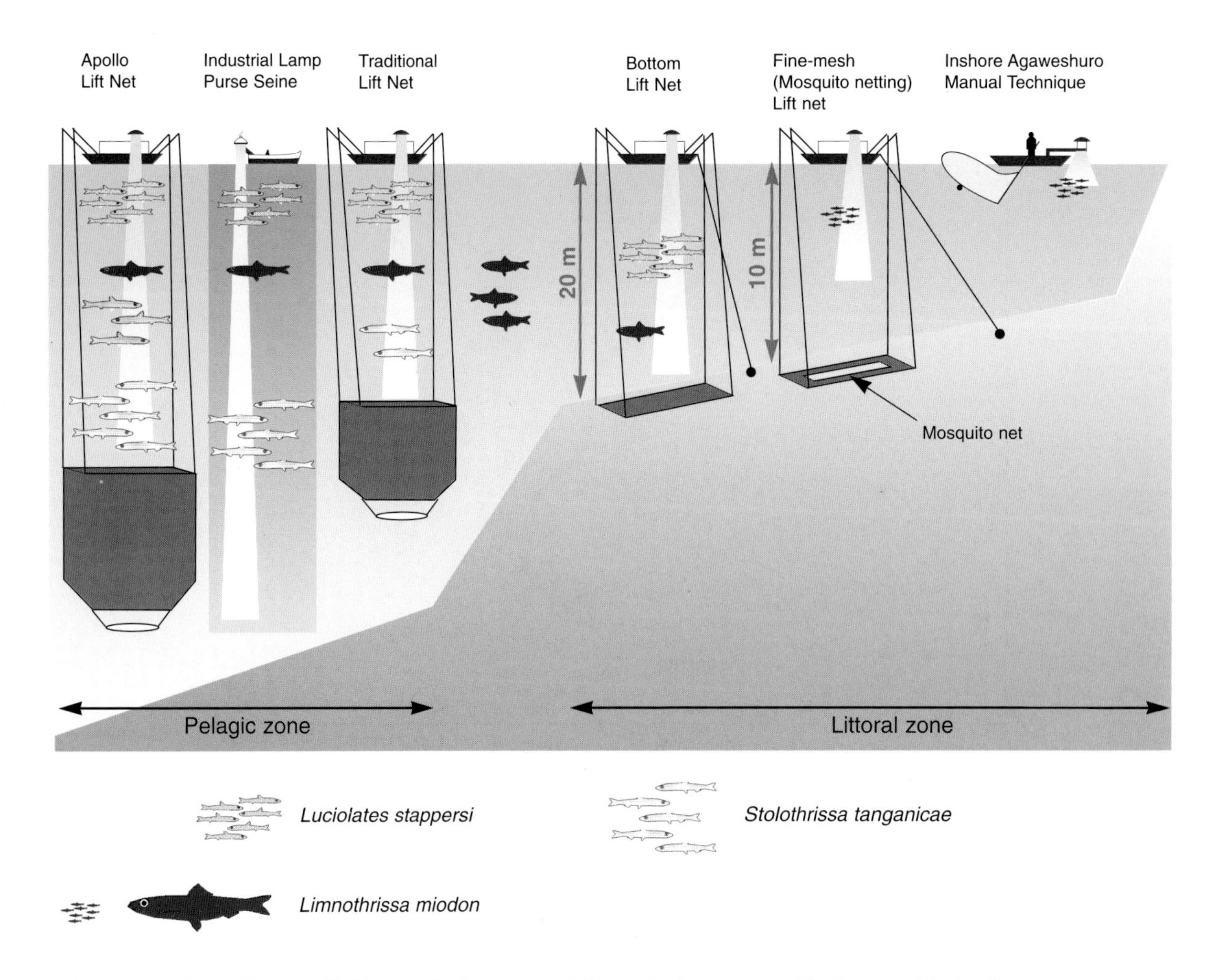

Figure 1.5 The different fishing techniques used for pelagic species (Petit, unpublished)

the sediment load in rivers such as the Malagarasi to a considerable extent which, in turn, will have an impact on the living communities (World Bank/DANIDA, 1994).

1.3 CICHLID FISHES

1.3.1 Taxonomy

Existing data

Cichlid species are widely distributed over the African continent and there are also many representatives in South and Central America. The number of cichlid species in Asia is small. Simple as it is to recognize a fish with only a single nostril on each side of the head and an interrupted lateral line system to be a cichlid fish, it is more difficult to recognize cichlids at the species level.

In Africa there are around 3000 freshwater fish species of which 80% are endemic to the Zaire system (Lowe-McConnell, 1993). The number of described cichlid species on the whole African continent is 870, as given in the *Checklist of the Freshwater Fishes of Africa* (Daget *et al.*, 1991), while an unknown number of species still awaits description. Some of these undescribed species have become extinct already (as in the case of a few hundred cichlid species from Lake Victoria), now only to be found in museum collections, or have disappeared unnoticed and unknown in number (as in the case of the Lake Kyoga cichlid fauna).

Although it is difficult to find any water body in Africa south of the Sahara without a cichlid fish species, very high numbers of cichlids are found in the three Great African Lakes. The estimated number of cichlid species according to Lowe-McConnell (1994), for the three lakes is: Lake Victoria, 300; Lake Tanganyika, 200; Lake Malawi, 500. This large number is believed to be the result of adaptive radiation, a fast evolutionary process in which a large number of species develops from a single ancestor.

The number of known cichlid fishes in Lake Tanganyika was estimated at the beginning of the 19th century to be 79 species, of which Boulenger (1905) described 60. Poll (1956) found 127 species of which he (re)described 55. In 1986, Poll reviewed the existing taxa and made a new list of 172 Lake Tanganyika cichlid species (Poll, 1986). This work by Poll was possible following a sampling programme undertaken during the Belgian Hydrobiological Expedition (1946–47), during which pioneering work on algae (L. Van Meel) and molluscs (E. Leloup) was also undertaken. These descriptions of cichlid fishes are based on specimens which were available at that time for taxonomic research in the European museum collections, of which the museums in Tervuren, Belgium (Poll) and London (Boulenger) were the most important.

Poll's descriptions are accompanied by information on reproduction, distribution, habitat, economic importance, catching techniques and so on, as far as the information was available for these aspects.

Since Poll's studies, several new nominal species have been discovered. An important role has been played by the numerous aquarists who have indicated more taxa than have been officially described. Aquarists have provisionally described these species in the periodicals of the cichlid ornamental fish groups. Approximately 50 other additional species have been recognized by Belgian taxonomists who are at present working on the taxonomy of Lake Tanganyika cichlids (J. Snoeks, Royal Museum of Central Africa and E. Verheyen, Royal Belgian Institute of Natural Science, personal communication). This increase in total species number can be mainly attributed to the division of 'problematic' existing taxa by means of mtDNA techniques to indicate and define taxa which are then re-examined for morphological characteristics.

Detailed knowledge on the taxonomy, speciation process and distribution pattern of cichlids is small, for the reasons listed below (Snoeks *et al.*, 1994).

- A limited number of specimens is available in museum collections, e.g. *Neolamprologus kendalli* and *N. 'cygnus-walteri'* are known in the Tervuren museum from single specimens (and thus a single locality) only.

- No comparative studies have been made on related species and little type material has been examined in depth. No effort has been made to review the existing collections, so erroneous data remains published in the literature.

- Many areas of Lake Tanganyika are relatively under-explored, especially the western part of the lake, and the Zairian shore is unknown territory. *Lamprologus meleagris*, for example, is only known from several specimens from one locality, Mwerasi, on the Zairian coast. The benthic fish community, because of the inaccessibility of its habitat, is also under-explored. It should be noted that Lake Tanganyika has a coastline of around 2000 km, and that it would take many years to investigate every rocky outcrop, beach or other habitat at close quarters.

- Different colour morphs within allopatric populations have been identified for a number of species. However, with present knowledge it is difficult to judge these different taxa at a specific, subspecific or infra-subspecific level. A good example may be the circumlacustrine taxa of the genus *Tropheus*, which consists of seven nominal species. A large number of different colour morphs and great variation in the number of dorsal and anal spines have been consistently noted in a number of isolated populations.

As the systematic knowledge of Lake Tanganyika cichlids is still developing, it is to be expected that identifications of cichlid fishes from Lake Tanganyika may contain errors. Snoeks *et al.* (1994) give an example of such a misidentification in an ecological study by Hori (1983). A handicap is the lack of a handy field guide depicting all the species and describing in brief the distinguishing field characters of each species. To date, only the voluminous publications of Poll (1956) and Brichard (1989) have been available. The illustrated books of aquarists such as Konings (1988) and Konings and Dieckhoff (1992) are often used to identify the species, despite their limitations. FAO's *Field Guide to the Freshwater Fishes of Tanzania* (Eccles, 1992) is not a useful guide for the cichlid species, as it details only a handful.

Of all the Tanganyika cichlid species known at present, only five are non-endemic. However, these non-endemics are only (or most commonly) found near inflowing river systems. This means that over 97% of all cichlid species in Lake Tanganyika are endemic to the lake. Recently it was noted that the exotic tilapia *Oreochromis leucostictus* had entered Lake Tanganyika, possibly via the Ruzizi river and Lake Kivu, where it was found prior to its discovery at the mouth of the Ruzizi (L. De Vos, Centre de Recherche en Hydrobiologie Appliquée, personal communication).

Ongoing research

At present three groups of Belgian institutes are engaged in taxonomic research on Lake Tanganyika cichlids. They are the leading experts on this subject and are:

- the Royal Museum of Central Africa (RMCA), Dr J Snoeks, Tervuren, Belgium

- the Royal Belgian Institute of Natural Science (RBINSc), Dr E Verheyen, Brussels, Belgium

- Centre de Recherche en Hydrobiologie Appliquée (CRRHA), Dr L. de Vos (who is also attached to RMCA), Bujumbura, Burundi.

The first two groups are involved in the taxonomy of lacustrine cichlids of Lake Tanganyika, and three expeditions have been made in recent years: 1991, 1992 and 1995. During these expeditions almost the entire Burundian, Tanzanian and Zambian coastlines have been sampled at approximately 10-km intervals. A 870-km length of coast comprising 81 localities, mainly in the rocky and sand/rocky littoral habitat (1–5 m deep) has been sampled. It is likely that this 'round-lakc' survey is the first attempt at making a systematic inventory of the fishes of Lake Tanganyika.

This taxonomic work aims to clarify the status of different taxa by means of mtDNA techniques, particularly for the problematic taxa: Eretmodini, Ectodini, Tropheini, Lamprologini and Haplochromini. The divisions in taxa found are expected to be related to morphological differences, as already indicated by some analyses (E. Verheyen and J. Snoeks, personal communication). Some of this work is being carried out in collaboration with Dr A. Meyer of the State University of New York, USA and Dr C. Sturmbauer of the University of Innsbruck, Austria.

The third group (CRRHA) is active in the inflowing rivers of the Lake Tanganyika basin. An extensive inventory of fishes of the Ruzizi river system, including Lake Kivu, has been made and a publication on the ichthyofauna of the Ruzizi River is in preparation by L. De Vos and L. Risch. A 3-year project to compile an inventory of the fishes of the Malagarasi River system was scheduled to finish in 1996. A publication on the icthyofauna of this river system is also planned. Both studies cover the entire fish fauna including the cichlid fish community.

One group from Liverpool, UK (Wheeler and Oakland of John Moores University) recently set up a survey on the Zairian coast. A total of 34 sampling sites were selected, mainly to investigate ecological and behavioural differences in the phenotypes of fishes of the eretmodine cichlids. However, because of difficult working conditions ashore this work has had to be abandoned.

Cichlid fish literature for Lake Tanganyika is based on the work of very few pioneers who conducted (or are still conducting) field observations on the lake. Despite this work, many aspects of our knowledge of these cichlids are repeatedly cited but are not actually based on direct observations or conclusions from scientific research.

1.3.2 Distribution

Geographical patterns

As far as the existing data permit, the geographical distribution pattern of cichlid fishes in Lake Tanganyika may be classified as follows:

- circumlacustrine, species which are found throughout the lake, e.g. *Boulengerochromis microlepis*;
- interrupted circumlacustrine, species which have a discontinuous distribution along the coastline of the lake, e.g. *Lamprologus ocellatus*;
- northern basin, species found along the shores of the northern basin where deep water is adjacent to the shoreline, e.g. *Neolamprologus longior*;
- northern tip, species only found in the shallow area north of the northern basin, e.g. *Neolamprologus pleuromaculatus*;
- southern basin, species found along the shores of the southern basin where deep water is adjacent to the present shoreline, e.g. *Neolamprologus leloupi*;
- southern tip, species only found in the shallow area south of the southern basin, e.g. *Altolamprologus calvus*;
- single locality, species known from one single locality, e.g. *Lamprologus stappersi*;
- multiple isolated localities, species found only in a limited number of isolated spots, e.g. *Lamprologus signatus*;
- deep benthic, species which are found in the deep shelf area above the deoxygenated water level, e.g. *Xenotilapia sima*.

No cichlids are known to live in the pelagic habitat of Lake Tanganyika.

The geographical distribution pattern for a number of species appears to divide into two or three main basin communities, and may be the reflection of a historic event in which a lower lake level induced allopatric speciation. According to this theory, the present boundaries between species are a reflection of historical allopatric divisions (Sturmbauer and Meyer, 1992). Distribution boundaries at present may also be the result of existing physical barriers, for example a sedimentation area of a lake tributary stream, which prevents a rock-bound cichlid reaching a nearby rocky habitat. It must be noted that this pattern is not found in other organisms such as gastropod snails (Michel *et al.*, 1992).

The existence of colour morphs makes the distribution pattern more complex, as numerous morphologically identical but differently coloured taxa can exist in one species without overlapping.

Habitat

The distribution of cichlid species is also classified according to habitat where the species are found (see section 1.2). These habitats can be divided into:

- the upper beach zone, the very shallow area where the waves break on sandy beaches;
- the littoral, the upper zone to a depth of approximately 20 m;
- the sublittoral zone, the zone below a depth of 20 m;
- the benthic, the more or less flat bottomed shelf area;
- the bathypelagic area, the deep but still oxygenated waters to a depth of 80 m in the north and 240 m in the south of the lake.

Often, authors do not define what they regard as littoral and sublittoral or inshore and offshore habitats.

Cichlids also can be segregated according to substrate type, as outlined in section 1.2, including:

- rocky
- sandy
- swampy
- sedimentation areas of inflowing rivers
- shell deposit areas.

A peculiar distribution pattern has been found in 15 species of Lamprologine cichlids, which have their breeding sites in the old shells of gastropods, mainly *Neothauma tanganicense* (Yanagisawa, 1995). Extensive shell beds are found especially in the deeper areas (Cohen, 1989). These 'shell graveyards' contain hardly a living specimen of the *Paramelania* and *Neothauma* species. The presence of shell graveyards strongly influences the distribution pattern of the shell-breeding cichlid fish species.

The benthic and bathypelagic fish community is dominated by cichlids but, contrary to the stenotopic cichlids of the sand and rocky littoral habitat, demersal cichlids are found at depths ranging from 20 m to 240 m (Eccles, 1986; Coulter, 1991a; Hori, 1995).

Abundance and stock size

It is generally assumed that the number of fishes is unevenly distributed over Lake Tanganyika. Cohen *et al.* (1993) found large differences in species richness between undisturbed, moderately disturbed and highly disturbed areas of excess sedimentation, with disturbed areas being less species rich.

A geographic area of high fish species numbers (hot spot) is generally assumed to be the most southern (Zambian) part of the lake. A good indication that this phenomenon is true, based on the number of Lamprologine species, is shown by van Wijngaarden (1995). It is possible that more of these hot spots may be located in Lake Tanganyika.

The environmental mosaic is of particular significance for the cichlids. An area with a number of different habitats, each supporting its own cichlid fauna, is likely to have a high number of species also. The highest species density of cichlid fishes is found on the rocky shores in the littoral habitat.

To quantify this, the cichlids in an experimental quadrat of 20 m^2 were counted on a number of occasions. In a littoral rocky area at Luhanga in Republic of Congo, 38 species with 7000 fishes were counted, i.e. 17–18 fishes per m^2. Most abundant in this survey were plankton feeders which comprised 56% of the total sample, but from only two species; omnivores (21% from seven species); aufwuchs eaters (18% with 15 species); zoobenthos feeders (eight species); and piscivores and scale eaters (4%, six species) (Hori, 1983). In a similarly sized littoral rocky habitat at the Mahale mountains in Tanzania, 44 cichlid species with 5700 fishes were counted, resulting in 14–15 fishes per m^2 (Hori, 1987). In a similar survey in the Zambian sector of the lake, 47 cichlid species were found with a density of 6–7 fishes per m^2 (Hori *et al.*, 1989). Although only three sites were investigated, it appears that the number of species may be negatively related to the abundance of fish.

Census results over a period of 10 years at the Zairian sampling site at Luhanga have also provided data on the persistence of the community structure. Although the number of fishes decreased by half over the period, partly due to a drop in lake level, the number of species at the site remained stable. A number of species disappeared, while others appeared (Hori *et al.*, 1993). In an additional experiment, 70% of the initially counted fish from an area of 70 m^2 were removed. Within 2 weeks the number of fish had recovered, while after 10 months the number of species at the location had recovered. The invading fishes were adults and sub-adults from adjacent areas (Hori *et al.*, 1993). It was concluded that the littoral fish communities in Lake Tanganyika are very stable and in a state of equilibrium, indicating that some mechanism or process exists to maintain the community structure.

Hori (1983) states that in special habitats Lamprologines form 75% of the total number of cichlid fishes, while elsewhere they form 25%. Lamprologines (formerly the genus *Lamprologus*) are, however, also one of the most species-rich taxa in Lake Tanganyika.

1.3.3 Ecology

General

It is remarkable that although the cichlid species of Lake Tanganyika are known to possess special features in their ecology, only limited documentation on the ecology of these fishes is available. Extensive literature concerning ecological aspects can be found in aquarists' journals. It would be incorrect, however, uncritically to extrapolate this information for cichlids in the field: these publications are written for a non-scientific audience. Nevertheless, it is worth referring to this literature for comparison with field results. Several aquarists have not only an extensive knowledge of all kinds of behavioural aspects from tanks, but also extensive field experience in Lake Tanganyika itself (Konings and Dieckhoff, 1992).

Many ecological studies have been conducted on Lake Tanganyika by a team of Japanese biologists in co-operation with Zairian biologists, under the guidance of Professor H. Kawanabe of Kyoto University, Japan. Pioneering work in many different fields, including systematics, genetic relationships, morphology, reproductive habits, food and feeding habits, population studies, interspecific relationships and parasitic studies on fish, has been carried out at sites in Republic of Congo, Tanzania and Zambia. Further studies have been carried out on the plankton, flora and fauna, benthic communities, relationships between fish larvae and shells, ecoclimatology, fishery statistics and bird behaviour as well as ethno-ichthyology.

Provisional research results have been published as an abstract in the *Ecological and Limnological Study on Lake Tanganyika and its Adjacent Regions*: by the end of 1995 nine of these reports had been published in Japanese (Kawanabe, 1995), some also available in English. Several reports are based on observations of the more peculiar aspects of cichlid life, for example, parasitic mouthbrooding, and have an anecdotal style (Sato, 1986). A disadvantage of many of these studies is that the research covers a limited period, often only some months and rarely over a year.

The key question in the ecological research on Lake Tanganyika cichlid fishes is how so many different species can coexist in limited areas, feeding on a limited food supply. Superficially it seems that all cichlid species depending on one food source share this source equally. This, however, has been shown not to be the case for aufwuchs-eating cichlids (Takamura, 1983b, 1984) and several species of the *Lamprologus* group (Hori, 1983). Both studies demonstrated that each species has its own unique hunting behaviour, method of feeding, type of food and so on.

Trophic types

Cichlids, like all other fishes, may be classified according to food type (e.g. algal feeders and piscivores) and method of feeding (e.g., algal grazers, browsers, scrapers and tappers). Twelve trophic types in cichlid fishes from Lake Tanganyika have been recognized by Hori *et al.* (1993), while Konings and Dieckhoff (1992) recognized nine groups containing a total of 22 trophic types as follows:

1: (1) aquatic weed eater

2: (2) algae scraper
(3) algae comber
(4) algae nibbler
(5) algae scooper

3: (6) detritus feeder

4: (7) zooplankton picker
(8) zooplankton sucker
(9) zooplankton filterer

5: (10) invertebrate browser
(11) invertebrate hunter
(12) invertebrate sifter
(13) invertebrate sonar-picker

6: (14) snail crusher
(15) snail sucker

7: (16) piscivorous hunter
(17) piscivorous stalker
(18) piscivorous sucker
(19) piscivorous ambush hunter
(20) opportunistic piscivore

8: (21) scale eater

9: (22) egg eater

No real pedophagous species are known, while predation of eggs is known. This is remarkable as in Lake Kivu (regarded as 'cichlid species-poor'), which drains into Lake Tanganyika, one such pedophage exists: *Haplochromis paukidens*. As far as is known there are no species which depend entirely on phytoplankton. Some authors state that sponges are not eaten (Konings and Dieckhoff, 1992) while others give them as food items (Hori, 1993).

A particularly interesting trophic group are the cichlids which feed on gastropod snails. These fish may be found in several African lakes, for example Lake Victoria where they are common. A small number of cichlids from Lake Tanganyika are also known to feed on snails. Lake Tanganyika harbours a number of snails with heavily calcified shells developed as a result of coevolution with predatory crabs, which have strong chela (West and Cohen, 1991). Nothing is known about the relationship between calcified snails and the snail-crushing cichlid fishes.

Conflicting information is found on the ecomorphological aspect of food and thick-lobed lips in *Lobochilotus labiatus*, which according to Hori (1987) feeds on insects, while West and Cohen (1994) indicate the crushing of snails with pharyngeal teeth as a specialization of this species.

Although all cichlid species have adaptations for a particular method of feeding, opportunistic feeding is common and all fishes will prey on unguarded eggs or a passing shrimp, for example.

Breeding

Cichlid fishes perform a complicated courtship ritual and the number of available breeding sites is often a crucial factor in the abundance of a species. Parental care for eggs and juveniles can last for several months (Kuwamura, 1988), while the breeding time for *Cyphotilapia frontosa* lasts 54 days (Balon, 1981).

The size of territory of the two aufwuchs-eating fishes, *Tropheus moorii* and *Petrochromis trewavasae*, varied from shallow waters, where such a territory had an average size of 1.59 m^2, to deep waters with an average territorial area of 24.15 m^2 (Kohda, 1987).

In Lake Tanganyika 57 mouthbrooding species have been recorded of which six are species with biparental mouthbrooding. There are 53 substrate spawners, and 21 species with a variety of other strategies. Remarkable strategies occur in several species of the Lamprologines, where 'helper systems', 'collective breeding' and 'harem systems' are found (Taborsky, 1984). Predation of eggs and larvae is common, although no specialized pedophagous species are known. Breeding parasitism is known from a catfish which deposits its eggs for incubation into the mouths of a number of cichlid species.

Males of a maternal mouthbrooder, *Tropheus moorii*, defend a territory for at least 5 months for courtship and feeding purposes, while the female leaves for other territory after the eggs are fertilized (Yanagisawa and Nishida, 1991). Brood-mixing eggs of different fishes, conspecific and heterospecific, is common among cichlids and is also found in Lake Tanganyika (Ochi *et al.*, 1995).

While mouthbrooding, most cichlids do not feed. However, there are a few exceptions in which the parent may feed itself or the juveniles, as has been found in Lake Tanganyika for *Cyphotilapia frontosa* (Yanagisawa and Ochi, 1991) and two species of *Tropheus*, *T. moorii* and *T. duboisi* (Yanagisawa and Sato, 1990).

1.3.4 Exploitation of cichlid fishes

Artisanal fishing practices

Although much research has focused on the pelagic fisheries of Lake Tanganyika, hardly any attention has been dedicated to inshore fishing activity in the lake (Roest, 1992). Little is known of the extent of the fisheries along the shores of Lake Tanganyika, although part of the life cycle of pelagic fishes occurs in inshore waters (F. C. Roest, personal communication).

The main target species for artisanal fishing in Lake Tanganyika are the pelagic species, which are only partly harvested at open water sites. A number of the fishermen, almost all in the southern part of the lake, attract the fish by light and move these lights inshore. At the inshore sandy beach areas, a fine-meshed beach seine is used to catch these fishes. This technique has a considerable bycatch of small (juvenile) cichlid fishes, which are caught from the beach where the net is pulled ashore (Ndaro,

1992). Experimental beach seining on the Tanzanian coast near Kigoma during daylight hours showed a high percentage of cichlids both in number and weight, 68–99% and 33–96%, respectively (Ndaro, 1990).

Commercial fishing activities targeting cichlid fishes are quite common. The main target species are those which grow to a large size, the most significant example of which is *Boulengerochromis microlepis*, known locally as *kuhe*. Although most cichlid species are small, a number of species can grow to sizes over 20 cm, and they are also exploited locally (Bayona, 1991a, b). The number of these large cichlid fishes is small, as they can be found only in a narrow belt along the lake shore. Fisheries of these large cichlid species are for this reason always on a small scale.

Although fishing on a small scale, the number of fishermen involved over the entire coast of Lake Tanganyika may be considerable, and may involve a variety of different fishing gear, e.g. gill nets, beach seines, hook and line, traps and hand lines (angling). With open access to the fisheries and an increasing human population, it is likely that more and more non-professional or part-time fishermen, usually small-scale farmers, will make a small investment in fisheries and, with a limited amount of fishing gear, provide themselves with sufficient fish to supplement their daily protein needs. This was observed during 1995 in the southern part of the lake (A. S. Cohen, personal communication).

Cichlids in the ornamental fish trade

Cichlid fishes have been collected and exported for the ornamental fish trade since the 1960s, when international airline companies increased the number of destinations in Africa and improved their services at the airports. The small-sized cichlids of the African Great Lakes became very popular among aquarists because of their small size and species diversity. Lake Tanganyika cichlids have the advantage that many colour forms exist within one 'species', while many species display peculiar behaviour patterns. This makes them more attractive than fishes from other origins to a large group of cichlid-specialist aquarists.

In the Lake Tanganyika area at least eight commercial collectors/exporters are active. Four of these are long-established companies owned by expatriates with a sound knowledge of the market and the fishes (Brichard, 1989). An irregular number of exporters consists of raiders who emerge from time to time and then disappear. The quality of the fish from these exporters is often extremely poor, according to one main importer into Germany and the Netherlands, many fishes being dead on arrival in the importing countries (Verduin, personal communication).

The export companies depend upon reliable connections with international airports, of which Bujumbura (Burundi) is the most important. Commercial traders in ornamental fish in East Africa usually organize the collection of these fish themselves, rather than buying specimens from local fishermen. At present, there are two companies in Bujumbura, one in Kigoma, two near Kipili in Tanzania and four in Zambia, while in Republic of Congo three exporters appear to be active, one in Kalemie and two in the Zairian part of the lake close to the border with Burundi. In addition, numerous amateur aquarists visit the lake to see the fish in their natural habitat and often return carrying 'small' collections to Europe or the USA. The overwhelming majority of cichlid fishes in the commercial ornamental trade are, however, bred in captivity in Europe and the USA. Most cichlid species can be bred in captivity and this is much cheaper than collecting them from the wild and transporting them from Africa. However, the need for a supply of wild-caught Tanganyika cichlids will continue, based upon:

- the import of unknown species and colour forms;
- maintaining a pure line of existing stocks in captivity, as interbreeding with other species can occur and occasionally a pure strain may disappear completely in the commercial breeding trade;

- fashion, which demands special species or colour forms – as many Tanganyika cichlid species and colour forms have an extremely restricted distribution, prices for these rare cichlids are high and collectors will not reveal details of the locations from which they were collected, thus it is possible that a particular species or colour form from the lake may be threatened, although thus far this seems not to have occurred with any species in Lake Tanganyika;

- preference of the market for wild-collected fishes.

The exact number of fishes and fish species exported from Lake Tanganyika is not known but, according to sources within the organizations of cichlid aquarists, an estimated 100 000 specimens are needed annually to operate an ornamental fish export company and ensure its financially viability.

A practice which occasionally occurs among fish collectors is to release fishes collected at remote or distant stations into sites within easy reach of the collector, in the hope that the fishes released will be able to colonize the area and create a stock of fishes for the exclusive use of the collector. Experiments in Lake Malawi proved that this method works well, but it is in essence a form of introduction.

Although the ornamental fish trade does not appear to have threatened the diversity of the Lake Tanganyika cichlids up to now, there seems to be no clear picture of which and how many fishes and fish species are leaving the riparian countries. In Europe there is usually no limitation on the import of ornamental fish species not listed by the Convention on International Trade in Endangered Species (CITES). At present, however, no Lake Tanganyika fish species is recorded on the CITES list.

European countries do not record either the number or the species of fishes imported. In the Netherlands and Germany the weight of the whole consignment including water is recorded (Woeltje, 1995). Proper recording and monitoring of the export would be useful. The effects of selective collection of fish species from the littoral community have not yet been studied.

1.3.5 Threats to the diversity of cichlid species

The existing cichlid species flock has developed in Lake Tanganyika over thousands of years. Most probably species have developed while other taxa have vanished in a natural process of succession. It is also likely that the human population bordering the lake used the fish stock as source of food, while human waste was disposed of in the lake. These factors appear to have had no dramatic effects on the number of cichlid species in the lake until recently. Increased human population pressure, however, in turn increases pressure on the natural environment including the aquatic habitat.

Threats to the abundance of cichlid species in Lake Tanganyika are:

- water pollution
- eutrophication
- destructive fishing
- species introduction
- siltation.

Water pollution and eutrophication are not specific threats to cichlids but will affect all kinds of organisms in the lake. Destructive fishing can have an impact on the number of cichlid species. However, the effects of fishing on the cichlid stocks of Lake Tanganyika are not clear. Species introduction can have a dramatic effect on a cichlid community, as has been observed in Lake Victoria (Witte *et al.*, 1992). However, although there are no known plans for introductions in Lake Tanganyika, the introduction of exotic species is difficult to avoid, as it could be effected by a single man with a bucket.

Siltation is a recently noted and extensive threat to the Lake Tanganyika cichlid fish community, although it is not a new phenomenon in the lake. The Malagarasi and Ruzizi rivers have developed large delta areas in the lake due to sedimentation. The number of sedimentation areas and speed of the siltation have increased (Cohen *et al.*, 1993). The effect of this sedimentation is to destroy the habitat of those cichlids which are dependent upon rocky and sandy shores. The habitat is simply buried by sediment. As the crevices and holes in the rocks where many cichlids find shelter against predation and where eggs are hidden disappear, so do the cichlid fishes of this community. This is particularly true for the species found in the rich rocky habitat. Algal feeders will disappear as algae have no substrate to grow on, while the light penetration will be reduced due to suspended matter. These effects will clearly have an impact upon species richness, but the extent and longer term implications remain to be determined.

1.3.6 Reserves, proposed reserves and research areas

In the Lake Tanganyika basin there are four national parks and one forest reserve with protected status that are adjacent to the lake. Ten locations have been suggested as aquatic parks and sites for research (Cohen, 1991). The present national parks received their protected status in order to conserve mammal and bird life: underwater fauna, however, is not specifically included. As the serious threats to a rich aquatic fauna in the littoral habitat derive from landside human influence, it may be possible to locate underwater national parks adjacent to these land-based national parks, or at least to incorporate some of the vulnerable catchment areas within the protected zone to safeguard their continuing existence (Coulter and Mubamba, 1993; Cohen, 1994). Potential areas could include the following.

- Ruzizi Delta, Burundi: inflowing river delta at the northern tip of the lake. Due to its size a peculiar habitat. Probably not very attractive for the biodiversity of cichlids due to low numbers of widely distributed species. Area well investigated by the recent research of L. De Vos and the Japanese team (see section 1.3.3).

- Gombe Stream National Park, Tanzania: undisturbed forested area drained by five small streams. Because of its location on a deep-water rocky shore, a potentially good representative site of the northern basin area.

- Mahale Mountains National Park, Tanzania: forested mountainous area on the northern side of the Kungwe mountains. The park includes a strip of coastline extending 1 km offshore into the lake. Because of its steep rocky shore interrupted by beaches, a diverse fish fauna is expected and this is indicated in a provisional inventory by Takamura (1983a).

- Nsumbu National Park, Zambia: located at the southern end of the lake and adjacent to a benthopelagic habitat. The protected area is a coastal strip of land 53 km long and extending 1.6 km into the lake. Fishing is permitted for 3 months of the year. The distribution of several cichlid species is restricted to this area and it is likely that a high degree of intralacustrine endemism may be present.

- Kigwena forest reserve, Burundi: located on the east coast of the northern basin and adjacent to a 10-km stretch of rocky coastline. Possibly a species-rich littoral rocky habitat.

- Malagarasi river mouth, Tanzania: this area includes the most extensive swamps and wetlands bordering the lake, and is thus already a special habitat of the lake. Numbers of cichlid fishes are low. The Malagarasi river is sparsely investigated because of its large size (L. De Vos, personal communication) and contains several endemic cichlid species.

- Luhanga-Pemba coastal strip, Republic of Congo: a steep rocky habitat interrupted by sandy beaches in the northern basin of the lake. Due to the research of the Japanese–Zairian team the cichlid fauna of this area is well known.

- Ubwari peninsula, Republic of Congo: steep rocky habitat interrupted by sandy beaches and site of Japanese–Zairian research team. A species inventory has been made by the Japanese team (Takahashi *et al.*, 1995).

- Moba-Moliro coastline, Republic of Congo: large stretch of coastline with deep cliffs adjacent to the deepest part of the lake. The human population is small but growing. Unexplored area with a high potential for undisturbed fauna.

- Lukuga river, Lake Tanganyika outlet, Republic of Congo: special habitat as there is only one outlet of the lake. Interesting for riverine cichlid species.

- Yungu to Kalemie Stretch, Republic of Congo: steep shore area adjacent to the northern basin. Several islands with potential for rich cichlid fauna. Little-explored area.

- Marshland south of Bujumbura, Burundi: probably of lesser importance for cichlid fishes due to its low number of species.

- Northern Burundi coast, Burundi: a rocky shoreline area with a relatively high diversity in fish fauna.

- Rocky coast North of Nyanza Lac, Burundi: rocky shoreline area with particular habitat because of the presence of algal reefs, 10–30 m deep and 100–500 m offshore, with a rich cichlid community (Cohen, 1991).

- Chituta Bay and adjacent regions, Zambia: fauna more or less similar to Nsumbu Bay but sensitive to human disturbance due to the harbour at Mpulungu.

1.4 NON-CICHLID FISHES

1.4.1 Introduction

While Lake Tanganyika, along with the other East African Rift Valley Lakes, is noted for the diversity of its cichlid species flocks, the fisheries in Lake Tanganyika are dominated by non-cichlid species. This contrasts with fisheries in Lake Malawi and the situation in Lake Victoria prior to the introduction of *Lates nilotica*. More than 90% of recorded fish landings from Lake Tanganyika are from just six species (Bayona, 1988): two endemic pelagic clupeids, *Stolothrissa tanganicae* and *Limnothrissa miodon*, and four endemic centropomids of the genus *Lates* (sub-genus *Luciolates*). This gives the non-cichlids a particularly important role in maintaining the 'integrity' of the Lake Tanganyika ecosystem.

Twenty-one families of non-cichlid fish are represented in the Lake Tanganyika basin (De Vos and Snoeks, 1994). Of the 145 recorded species from 51 different genera, 61 species are endemic. Thus the diversity of non-cichlid fishes approaches that of the more celebrated cichlid fishes, although the number of recorded species of the latter may be a considerable under-estimate (172 species, 167 of which are endemic; 50 genera; Coulter, 1991b). The numbers of non-cichlid species and genera differ slightly from those reported by Coulter, as several genera were renamed in subsequent work, and several new species are being described (De Vos and Snoeks, 1994).

The origins and evolution of the fish communities are reviewed and discussed by Coulter (1991c). The known distribution of the non-cichlid fish in the Lake Tanganyika basin is outlined by De Vos and Snoeks (1994).

All species of the families Clupeidae and Centropomidae are endemic, and are either pelagic or deep-water benthic. Other families with high endemism are primarily benthic, and include the Bagridae, Mochokidae and Mastacembelidae. Families with no, or a lower proportion of, endemic species are those of riverine origin, although some of the species may be found within the lake, in the littoral and deep benthic zones.

1.4.2 Main groups of non-cichlid fish

Characterization

Kawanabe (1991) identified three levels of biodiversity in the fishes of Lake Tanganyika: within-species genetic diversity, taxonomic species diversity, and ecological diversity. The lake's fish communities are known to be rich in the latter two types, due to the breadth and diversity of environmental conditions within the lake, and its great age. The long evolutionary history and diversity of ecological niches have allowed diverse fish communities to evolve. The large size of the lake and limited mobility of many of the fish species have perhaps also led to high diversity of genotype in some species. Unlike the cichlids, no localized forms of non-cichlid are known within the lake; of the 65 species of littoral and benthic non-cichlids in the lake, 75% are known to have a circumlacustrine distribution (De Vos and Snoeks, 1994).

Taxonomic references and field identification keys

The major reference work on the taxonomy of the non-cichlid fish of the Lake Tanganyika basin is that of Poll (1953), who references the type material and provides detailed descriptions, key identification features and line drawings of many species, as well as notes on distribution, habitat and ecology. A dichotomous key to the entire non-cichlid fish fauna is also provided. This can be used even if one does not know the family of the fish. A revised list of genera for all African freshwater fish, together with identification features, is given by Poll (1957). A key to the families and genera of freshwater fish of Tanzania is given by Matthes (1975). River fish from the Tanganyika/Kivu region are described by Poll (1952).

Updates and corrections to these earlier works can be found through the *Checklists of the Freshwater Fishes of Africa* (Daget *et al.*, 1984, 1986a,b). The latter contains a complete bibliography of the taxonomic literature for non-cichlid African freshwater fishes. A similar volume is also available for the cichlids (Daget *et al.*, 1991).

A field guide to the fishes of Tanzania designed for species identification by fisheries personnel (Eccles, 1992) may also be found useful, although it covers all of Tanzania, not just the Tanganyika basin, and is confined to species of potential or existing fishery interest. Pictorial outlines and a guide to orders and families enable broad-scale classification, and keys and drawings then allow users to identify fish to species level. The key may be particularly useful for identifying some of the more widely distributed fish of riverine origin. A bibliography provides access to most of the original taxonomic work, and to some recent taxonomic revisions and new species descriptions. Several species are currently being described (unpublished data, reported by De Vos and Snoeks, 1994).

Brichard (1978, 1989) provides colour photographs of most species, which may be an aid to their field identification. He also provides keys to species within each family or genus. These are generally adapted from previous keys and descriptions, but several appear to be original. Some misidentified photographs are apparently included (Rossiter, 1993).

Most keys are based on the identification of adults, as the life histories of many species are not well known. Criteria for distinguishing juvenile *Lates* spp. have been studied by Kinoshita and Tshibangu (1989).

A field identification manual with computer-generated images of each fish species, together with a table of basic information on distribution, ecology, Latin and local names, has been prepared for the exploited species in the north (Burundi waters) of the lake (Petit and Nyakageni, 1994) and is being updated.

The comprehensive keys for field identification of non-cichlid fish are rather outdated, and the more modern work is either incomplete or too general to be used for rigorous identification to species level in biodiversity surveys. There is a requirement for a comprehensive dichotomous key (based on characteristics identifiable in the field) to the non-cichlid fish of the Lake Tanganyika basin, including the affluent and effluent rivers and streams.

1.4.3 Distribution and ecology of the pelagic fish

General characteristics

The pelagic fish community is composed of large populations of six endemic species: two zooplanktivorous clupeids and four predatory centropomids. Because the pelagic fish community supports important fisheries (Coulter, 1991d; Roest, 1992; Greboval *et al.*, 1994; Pearce, 1995), more is known about the population dynamics of this fish community than for any offshore fish community in the African lakes (Coulter, 1991d).

Clupeids

General characteristics

Clupeids are prominent in pelagic zones of many large lakes around the world, and are found in several of Africa's large lakes, both natural and artificial (e.g. Lakes Volta, Kainji and Mweru). Like the marine clupeids, their lacustrine counterparts tend to be small pelagic fish, numerous and short-lived, highly fecund and highly productive. Because of this, they have been considered prime candidates for introduction into new water bodies for the purpose of developing new fisheries: numbers of *Limnothrissa miodon*, from Lake Tanganyika, have successfully been introduced into Lake Kariba, where they support a productive pelagic fishery (Marshall, 1993). Introductions into Lake Kivu have been less successful, while the ultimate result of the accidental introduction into Cabora Bassa, and the deliberate introduction into Ithezi-tezi, must await the availability of data collected over a longer period.

Stolothrissa tanganicae

S. tanganicae become sexually mature at 75 mm total length in females and 64 mm in males (50% maturity). Sexually mature individuals occur throughout the year and individuals may spawn several times in a year. Different studies indicate variety in peak spawning seasons (Coulter, 1991c for review), but Pearce (1985a) suggests that the apparent marked reproduction periods are in fact determined mainly by seasonal variation in survival rates of eggs and fry. This is brought about by differences in timing of seasonal production cycles, with cohorts of fish from spawnings arising earlier in the year in the south of the lake than in the north, a pattern that follows the south–north progression of the period of maximal mixing (Coulter, 1991e). This interpretation is consistent with the findings of a recent detailed study on the larval ecology of the small pelagic cyprinid *Engraulicypris sardella* which fills a similar ecological niche in Lake Malawi (Thompson, 1995). In the case of Malawi, survival was determined by the effects of wind-induced mixing of nutrient-rich hypolimnetic water, giving seasonal increases in primary production that were strongly coupled with crustacean zooplankton production and larval fish production (Allison *et al.*, 1995).

S. tanganicae eggs double in size after fertilization and sink slowly (4–5 cm/min). Studies in the southern part of the lake indicate that hatching occurs 24–36 h after fertilization at 75–150 m depth, and the larvae are immediately capable of upward swimming movements to remain in oxygenated water (Matthes, 1967). The depth of the oxygenated zone is less in the northern part of the lake, presumably affecting larval survival. Natural mortality rates have been calculated by Moreau *et al.* (1991), using the empirical formula of Pauly (1980), as 3.5–5.4. Analysis of length frequency distributions give estimates of von Bertalanffy growth parameters of L_{inf} = 9–10.7 cm and k = 1.6–4.0, for various years and areas. Most population parameters show high variability between years, probably due to environmentally mediated density-dependent interactions.

The larvae, which are transparent, are thought to migrate vertically (Poll, 1953). They start to school at 10 mm total length, when silvery pigmentation becomes apparent. Schooling patterns and behaviour are reviewed in some detail by Coulter (1991e). The onset of schooling behaviour may correspond to the period when they become visible and vulnerable to piscivorous *Lates* species. Schools of juveniles move progressively nearer shore as they grow, and fish of 35–50 mm total length are the target of inshore scoop-net and beach-seine fisheries (Coulter, 1970). From 50 mm, they move offshore and begin to appear in the offshore fishery at 55 mm (FAO, 1978a). A second inshore migration is thought to occur for spawning, the evidence for this being provided by the absence of ripe-running fish in pelagic catches, and their common occurrence near shore during the main spawning season (Mann and Ngomiraakiza, 1973; Enderlein, 1976). Although both *L. miodon* and *S. tanganicae* are pelagic, each has an inshore phase in its life cycle, more strongly marked in *L. miodon*. A study in Republic of Congo waters (Mulimbwa and Shirakihara, 1994) has shown that *S. tanganicae* spawns offshore (>0.5 km), while the spawning site for *L. miodon* was not determined. *L. miodon* larvae off Uvira (Tshibangu and Kinoshita, 1989) were more common inshore, while *S. tanganicae* were most abundant 1.5–2.5 km offshore. Populations of fish larvae sampled 2 and 7 km off Uvira in 1987–88 indicated that 96% were clupeid larvae, with *Lates* making up the remainder (Tshibangu, 1988); no larvae were found in the Ruzizi river mouth. These represent the only direct studies of larvae to date; most other studies have relied on inferences made from fishery catches. Presumably the larvae then become dispersed offshore, until their inshore movement at 35 mm, although Coulter's review (1991e) does not explain this apparent paradox.

The lowest depth to which *S. tanganicae* are distributed is probably determined by oxygen concentration: they appear tolerant of levels of around 2–3 g O_2/m^3, which corresponds to depths of 180 m in the south and 100 m in the north of the lake. Vertical migrations take place mainly at dawn and dusk, with large, tight schools being formed in daylight, and the fish being dispersed in near-surface layers on nights with no moon. During full-moon periods, aggregation in sub-surface layers is fairly tight, with evidence of small schools visible (Coulter, 1991e). Upward vertical migrations at dusk are thought to occur in response to migration of their zooplankton food (Poll, 1953), while downward migrations at dawn occur to seek optimum light conditions and minimize predation risks (Coulter, 1961). Feeding takes place most intensively at dusk, but also throughout the day. Juveniles living inshore feed mainly on phytoplankton, with the diatoms *Nitzschia* and *Navicula* and peridinid *Gymnodinium* predominating in the stomachs sampled (Chéné, 1975). The switch to the adult diet of zooplankton seems to occur around the time of migration into the offshore pelagic zone, at 50 mm total length, with the dominant crustacean zooplankton, *Tropodiaptumus simplex*, being the main component of the diet (Chéné, 1975), and *Limnocaridina* being sporadically important (Marlier, 1957), probably being preferentially selected when encountered. A study of the diets of inshore *S. tanganicae* and *L. miodon* in the south of the lake (Phiri, 1991a) suggests that atyid prawns may be the most important food for these species. This study was, however, based on small samples taken at one locality over a 1-month period.

The evidence for extensive lateral movements of *S. tanganicae* within the lake is equivocal, and based mainly on acoustic observations of changes in stock biomass in particular areas (reviewed by Coulter, 1991e), but it is important for fisheries management purposes to establish whether flux is sufficient for the population to be considered a unit stock, or whether several local sub-stocks exist. The lags in peak spawning period between the north and south are evident from the appearance of cohorts of adults in the fishery, at different times and in different areas. This suggests that extensive mixing does not take place, and that *S. tanganicae* move relatively little. There is no evidence of morphologically differentiated sub-stocks, although random amplified polymorphic DNA analysis (RAPDNA) on small samples of *S. tanganicae* and *L. miodon* has indicated that genetic differences may occur between areas (Kuusipalo, 1994). This indicates that these two pelagic fish may not come from single homogeneous stocks, and has considerable management implications.

Limnothrissa miodon

L. miodon, like *S. tanganicae*, appears to reproduce year-round, but with the spawning taking place during the November–May rainy season, with a minor peak near the beginning and major peak near the end of this period (see Coulter, 1991e; Marshall, 1993 for reviews). Although spawning is also inshore, in contrast to *S. tanganicae*, *L. miodon* appears to be a bottom spawner, with eggs hatching on sandy substrates in waters less than 130 m deep (Matthes, 1967). The newly hatched larvae resemble those of *S. tanganicae* and commence schooling at a similar size. Large schools of juveniles, 15–40 mm long, are found close inshore (Figure 1.4), usually in June–July in the southern part of the lake (Pearce, 1985a). Schooling behaviour is similar to that of *S. tanganicae*. Size at maturity shows seasonal variation in females, from 64 mm in April to 83 mm in August (Ellis, 1971). A general offshore movement occurs at around 100 mm, usually in October–November. *Limnothrissa* generally have a 2-year life-span; fish over one year generally remain in the offshore pelagic zone, and are piscivorous on *S. tanganicae*. A biphasic growth pattern has been observed for *L. miodon* in Tanzania, with faster growth above 120 mm associated with a sudden switch to a piscivorous diet at around that size, associated with an offshore migration (FAO, 1978b). In Zambia, where movement offshore into deeper water is more gradual, a more typical growth curve, having decreasing increments with increasing size, is observed (Coulter, 1991e). A study of growth, reproduction and recruitment of two clupeid species off the Republic of Congo coast in the north-west of the lake has shown that *L. miodon* generally reach smaller maximum size (13.5 cm) than in Zambian and Tanzanian waters (15–16 cm) (Mulimbwa and Shirakihara, 1994). An overall total mortality rate of 0.483 per month has been calculated from length-frequency analyses, with total and natural mortality rates being difficult to separate (Coulter, 1991e, Table 6.III). Moreau *et al.* (1991) report natural mortality rates of 2.2–3.6/year, derived from Pauly's (1980) empirical model, and use the ELEFAN 1 routine on length-frequency distributions to calculate von Bertalanffy growth parameters of L_{inf}=16–17.5 cm and k=1.1–1.3 for various years and areas within the lake.

The growth rates of *S. tanganicae* and *L. miodon* from Zambian waters around Mpulungu were assessed from daily rings on otoliths (Kimura, 1995) and found to be similar to those derived from length-frequency analysis. The growth and breeding seasonality of clupeids has also been studied off Uvira (Mulimbwa, 1991; Mulimbwa and Shirakihara, 1994).

In general, *L. miodon* occupies a more inshore habitat throughout its life than *S. tanganicae*, with much of the population inhabiting the relatively shallow shelf areas, such as that off the Zambian coast, where they used to form over 50% of the clupeid catches in both inshore and offshore fisheries (Pearce, 1985a). *L. miodon* is much less important where the lake littoral shelves abruptly, e.g. at Kigoma (FAO, 1978a). Populations are generally thought to be more stable than *S. tanganicae* (Coulter, 1991e). This can be attributed to the greater range of food resources available to a larger, more inshore species – its survival is not so tied to fluctuations in offshore primary productivity as *S. tanganicae*. *L. miodon* is planktivorous when young, consuming phytoplankton, crustacean zooplankton and insect larvae (Poll, 1953; Matthes, 1967), and it becomes piscivorous on *S. tanganicae* when older: *L. miodon* of 58–115 mm length have been observed to feed on *S. tanganicae* of 27–65 mm length (Coulter, 1991e).

The extent of lateral movements within the lake is unknown, but preliminary RAPDNA analysis suggests the existence of discrete sub-stocks (Kuusipalo, 1994). Some preliminary work on stock genetics of the two clupeid species from the north of the lake has been reported by Nishida (1988).

Centropomids

General characteristics

The *Lates* species (sub-genus *Luciolates*) are among the most abundant and ecologically important fishes in the lake. *Lates mariae* is largely benthic, *L. angustrifrons* is an inshore predator, while *L. microlepis* is the top pelagic predator, feeding on the clupeids and to a lesser extent on its smaller pelagic congener *L. stappersi*. Although *L. mariae* and *L. angustifrons* are often found in the pelagic zone, they are more typically associated with demersal fish communities, and current knowledge on the ecology and distribution of these two species is reviewed in the section on the benthic fish (section 1.4.4).

The ecology of juvenile *Lates* species inhabiting reed beds has been studied by Kondo and Abe (1989) in the Mbemba region. *L. stappersi* is entirely pelagic, and not found in reed beds; *L. microlepis* is rare in this habitat, and its juvenile ecology not well known. *L. angustifrons* is dominant in short reeds while *L. mariae* is the predominant species in long reeds. Specimens of 7–119 mm were found in reed beds. It was considered that both *L. mariae* and *L. angustifrons* were nocturnal predators on shrimp, and to a lesser extent on fish fry. Shrimps are abundant in reed beds, with 12 of the 13 recorded endemic species being found in this habitat.

Lates (Luciolates) stappersi

L. stappersi may spawn 3–4 times a year, as evidenced by the presence of several size classes of eggs in each mature ovary (Pearce, 1985a). Total egg production per female was estimated as 10^4–10^6, or about 550 eggs/g body weight. Size at 50% maturity is 18–22 cm, with males maturing at a slightly smaller size (Coulter, 1991e, Table 6.IV). Outside the main breeding season (February–April), only fish larger than 30 cm were found in spawning condition (Ellis, 1971). *L. stappersi* has pelagic eggs (Pearce, 1985a; Coulter, 1991b), and post-larvae occur in the plankton (Poll, 1953). All *Lates* species are common in the plankton from several millimetres to 25 mm in size. Juvenile *L. stappersi* are distinguishable macroscopically from other *Lates* at sizes below 10 mm (Poll, 1953). They are pelagic, and pass most of their life cycle in the offshore pelagic zone, whereas the other *Lates* species have inshore juvenile phases. *L. stappersi* do have an inshore phase, in their second year at age 10–22 months (length 130–250 mm), associated with feeding activity on young and maturing stocks of *S. tanganicae*. *L. stappersi* recruit to the pelagic fishery at 200–290 mm (Roest, 1988).

Juvenile *L. stappersi* form schools that are associated with the main annual cohort of *S. tanganicae*, which hatch at the same time (Chapman and Van Well, 1974b). Pearce (1995), however, reports that the two species are seldom caught together in the south of the lake. Indeed, juvenile *L. stappersi* are rare, whereas they are common in the north (Roest, 1992). Adults concentrate in layers at 20–30 m deep at night (Ellis, 1971). Young *L. stappersi* feed on *Mesocyclops*, *Tropodiaptomus simplex*, and shrimps of the genus *Limnocaridina* (Chéné, 1975; Ellis, 1978) until they reach around 130 mm length, when they become increasingly piscivorous on *S. tanganicae*; shrimps remain an important or dominant prey item for adults, however (Chapman *et al.*, 1974; Pearce, 1985a), particularly in the southern part of the lake (Pearce, 1989, 1995), although *S. tanganicae* adults make up the majority of the diet by weight (Pearce, 1991a, b).

The juveniles of *L. stappersi* are preyed on both by adults of the same species and by *L. microlepis* (Coulter, 1991e).

Studies on the basic population dynamics of *L. stappersi* are summarized in Table 6.IV of Coulter (1991e). More recent studies by Roest (1988) and Moreau and Nyakageni (1992) have supplemented these findings. Maximum length (von Bertalanffy L_{inf}) varies from 45–47 cm, k=0.34–0.40. Total mortality rates between 1962 and 1983 have varied between 0.5 and 2.5 per year, and natural mortality rates have been calculated empirically from Pauly's (1980) equation as 0.67 per year (Moreau and Nyakageni, 1992). Longevity is 6–9 years.

Variations in growth and mortality rates, calculated from length-frequency analysis, have been quantified for *L. stappersi* from commercial purse-seine catches in the Burundi waters of the lake (Moreau and Nyakageni, 1992). Purse-seines are relatively unselective, with fish in the size range 6–50 cm total length being caught, providing data suitable for length-frequency analysis. Reproductive periods were shown to vary between years, making growth rate estimates from length-frequency distribution analysis difficult, as the age difference between cohorts was not always 1 year. Natural mortality rates were estimated at around 0.70 per year, with total mortality fluctuating from 1.6 to 2.5 between 1972 and 1983.

Lates microlepis

Juvenile *L. microlepis* are found in littoral weed patches, composed mainly of *Ceratophyllum*, *Vallisneria* and *Potamogeton*. They are also found around the roots of emergent vegetation (e.g. *Phragmites*) or even among rocks (Thompson *et al.*, 1977). *L. microlepis* appear to use weed beds as nursery areas, and may spend up to a year (25–180 mm total length) in these habitats. These habitats are sparse around the lake, and may be the 'critical factor' (*sensu* Hjort, 1914) in determining the size of the population. Young *L. microlepis* live inshore after leaving weed cover. They are recruited to the pelagic zone on maturity, at around 500 mm total length. They mature at age 3–4 years, and have a life span of around 10 years. L_{inf} has been estimated at 83–108 cm, with k=0.34. A summary of their population dynamics is given in Table 6.IV of Coulter (1991e).

L. microlepis is an active, surface-water predator of clupeids, common in areas where fishing has not been intense, and often closely associated with moving schools of *S. tanganicae*. It is thought to be more mobile than *L. stappersi*, ranging further in its foraging activities (Coulter, 1991e).

Cyprinidae

Chelaethiops minutus is the only one of the 46 described cyprinid species from the Tanganyika basin that occurs in offshore waters (De Vos and Snoeks, 1994). The young stages, up to 75 mm in length, occupy the littoral zone, while older specimens are found in the pelagic. They appear to feed from the water's surface on insects blown offshore from the land (Poll, 1953). However, this conclusion is based on limited samples, and bearing in mind the variability in diet of many fish species, it should be treated with caution.

The pelagic fish community

The pelagic clupeid and *Lates* populations form a closely interacting, co-evolved community, each species strongly influencing the ecology, behaviour and population dynamics of the others (Coulter, 1991e). The differing morphology of the southern, central and northern basins of the lake, spatial differences in the depth of the oxygenated layer, and the extent and timing of mixing events all affect the distribution, relative abundance and dynamics of the pelagic fish species. Variations in the distribution of fishing effort could have wide-ranging effects on the ecosystem as a whole. This influence is due to the important effects that the pelagic species have in structuring the pelagic ecosystem (top-down effects) and influencing the dynamics of inshore regions. The impact on the inshore habitats occurs either by predatory *Lates* restricting the cichlids to inshore habitats (Coulter, 1991b), or by the direct predatory or competitive effect of the presence of juvenile phases of many of the pelagic species in the inshore waters.

The biology of the pelagic fish species in the south of the lake is briefly reviewed, with reference to the fishery and the changes that have taken place in it, by Pearce (1995).

Trophic relationships in both the exploited and unexploited communities are reviewed and described qualitatively by Coulter (1991b). Quantitative models of the pelagic food web have been constructed using the ECOPATH software (Moreau *et al.*, 1993), while more general studies of pelagic ecosystem structure, function, productivity and efficiency of energy transfer are reviewed by Hecky (1984, 1991).

1.4.4 Distribution and ecology of benthic and bathypelagic fish

General characteristics

The benthic fish community in the deep waters of Lake Tanganyika is rich in endemic species and phylogenetically diverse. Its lower limit is determined by oxygen concentration. Eccles (1986) suggests that the lack of a de-oxygenated hypolimnion in Lake Victoria and the stability of the hypolimnion in Lake Malawi have allowed the evolution of a wide diversity of deep-water species restricted to narrow depth ranges in those lakes; while in Lake Tanganyika, upwelling of de-oxygenated hypolimnetic water causes great short-term changes in oxygen concentration over the depth range 50–250 m, so that fish species inhabiting this zone need the ability to adjust their depth range, and species diversity in this lake is thus lower. This appears to be a misconception: in total, 80 species are found in these habitats at depths greater than 40 m, and 78 of these are endemic. Seven families containing eight genera are represented. Following Coulter (1991f), the deep-water habitat can be conveniently divided into three categories:

- shelf areas, where the bottom slopes gently and is generally composed mainly of sand, mud and organic detritus;

- rocky slopes, where (as in much of the lake) the gradient is steep;
- the bathypelagic habitat.

Whereas no species in Lake Tanganyika was confined exclusively to depths below 60 m, in Lake Malawi at least 30 cichlids and three bagriid catfish are found exclusively below these depths. Deep-water species in Lake Tanganyika are partitioned by diet rather than by depth distribution.

With the exception of the Zambian waters of the lake, the demersal fishery has received little attention. In Zambia, extensive shelf areas and a deeper oxygenated layer provide a habitat for a community of demersal fish that are subject to exploitation by both subsistence and commercial (artisanal and industrial) fishers. Little is known about the ecology of the fish community in these habitats in comparison to levels of knowledge about the pelagic fish and the littoral cichlid fish communities. What is known comes mainly from bottom-trawl catches made by the Belgian Hydrobiological Expedition of 1946–47, and from gill-net catches in the south-east and south-west arms of the lake, in surveys conducted in 1961–63 (Coulter, 1966); in 1969–72 (Kendall, 1973a, b); and in 1979–83 (Pearce, 1985b, c). Gill-net fisheries were not established prior to the 1960s, so the results of the surveys by the Belgian expedition and by Coulter in 1961–63 can be considered indicative of the virgin fish community.

Non-cichlid fish dominated the gill-net catches by weight, with *L. mariae*, *L. angustifrons* and six species of *Chrysichthys* together making up over 60% of catches (Coulter, 1991f, Table 8.II). *Lates* were not significantly represented in trawl catches, however, and cichlids of the genus *Trematocara* were dominant (Poll, 1952; Coulter, 1991f, Table 8.II). Much of the benthic ecosystem in Lake Tanganyika is not suitable for trawling, being rocky, steeply shelving or consisting of very soft mud. Trawling may therefore not give the relatively unbiased samples of fish community composition that is normally assumed. The bias of gill nets towards active, predatory fish, even when a range of mesh sizes is used to reduce size selectivity, is well known (Hamley, 1975). Some of the differences between the two surveys, however, are attributable to the different substrates and areas sampled by the two gears. Trawling was lake-wide, but restricted to sandy substrates, while gill-netting covered a wider range of habitat types, but was restricted to the south of the lake. A close examination of the original data is recommended to try and disentangle the relative effects of gear, substrate and locality on the observed differences in fish species composition. These data could then be used as baseline studies for comparison with repeat surveys. Particularly useful is the species composition by depth data (e.g. Coulter, 1991f, Table 8.III) which could be used to decide how deep to extend the boundaries of any aquatic reserve. It should be noted that the maximum number of recorded species occurred at depths of 40–100 m, and that all species found below these depths were also found in the shallower areas, i.e. there appear to be no unique, exclusively deep-water species as there are in Lake Malawi (Eccles, 1986; Thompson *et al.*, 1995a,b). Depth segregation of age or size classes of some fish has been observed, however, and consideration should be given to the need for deep-water habitat by particular life-history stages of some species.

The bathypelagic fish community, living above the bottom in deep water, is caught neither by bottom-trawls nor by surface-gear such as lift-nets and purse-seines. It is possible that the abundance of fish in this area, which may extend over deep anoxic areas throughout the lake, has been underestimated by past surveys, as it had until recently in Lake Malawi (Thompson *et al.*, 1995 a,b), due to difficulties in sampling it. This was resolved in Lake Malawi by mid-water trawling, and by using 'gangs' of gill nets, set to drift at pre-determined depths (Allison *et al.*, 1994).

Coulter (1991f) suggests that the shelf fish community, relatively independent of the littoral zone, is sustained partly by the rain of organic material from the pelagic zone, and partly by fish feeding on pelagic prey. The main piscivorous species in the shelf fish community, *L. mariae*, *L. angustifrons* and *Dinotopterus cunningtoni*, all migrate into the pelagic zone to feed at dawn and dusk. The relative importance of the 'rain of detritus' and Vinogradov's 'ladder of vertical migration' as benthic–pelagic coupling mechanisms is an important area of study in deep-sea biology (Angel, 1990), and is of relevance to the component of this project studying energy flows. Studies of the deep-benthic and bathypelagic fish communities in Lake Malawi suggest that vertical migration of prey organisms from the euphotic zone is the most important means of energy transfer from the pelagic to the profundal zones (Allison *et al.*, 1995a; and in press).

Many of the benthic fishes are omnivores, but in contrast to shallow freshwater habitats in the tropics, few are detritivores. This may relate to the low organic content of the substrate in the deeper oxygenated zone (Coulter, 1991d). The detritus is generally utilized indirectly through feeding on invertebrates. The siluroids are active scavengers and feeders on macro-invertebrates, as are many of the cichlid species. The benthic food web is highly complicated, but amenable to modelling of trophic interactions and energy flows if mean trophic positions of the main organisms are established using carbon and nitrogen stable isotope-ratio analysis. The possibility that shrimps may play a similar ecological role to lake-fly larvae in Lake Malawi in sustaining the deep benthic and bathypelagic fish communities should be considered. The composition of the deep-water zooplankton is not well known (Coulter, 1991d). The extent and importance to the ecology of the littoral of lateral and vertical mass diurnal migrations of deep-water fish should also be considered.

The fish community on the shelf zones, although dominated by cichlids in terms of species numbers, supports larger numbers and higher biomass of non-cichlid fish, of which *Lates* and *Chrysichthys* species are the most important (Coulter, 1966). The rocky slope habitat, on the other hand, is dominated numerically by cichlids, mainly of small size, and of similar composition to the shallow, littoral rocky shore assemblage of species. Most of the predators are relatively small cichlids, but the siluroid *Chrysichythys grandis* is also common (Coulter, 1991f). The bathypelagic habitat is thought to support only zooplanktivorous cichlid species. Several riverine siluroids also occur in all three deep-water habitat types; these include *Auchenoglanis occidentalis*, *Malapterurus electricus* and *Synodontis nigromaculatus*. It is notable that while a number of deep-water species range into the littoral, especially at night, very few littoral species enter the deep benthic zone.

Centropomidae

While *L. microlepis* and *L. stappersi* are almost exclusively pelagic, the other two *Lates* species are found in the demersal fish community. *L. angustifrons* is the top predator in the demersal fish community and the largest species in the lake, and *L. mariae* is the most abundant demersal predator (Pearce, 1988). Adults of both *L. angustifrons* and *L. mariae* move into the pelagic zone for part of the year (Coulter, 1976) and play an important predatory role there. The three larger *Lates* species, *L. microlepis*, *L. mariae* and *L. angustifrons*, have declined in abundance wherever industrial fishing has taken place, while *L. stappersi* has generally become more abundant.

Juvenile *Lates* are found in reed and weed beds in the littoral zone for much of their first year of life, although post-larvae are found in the plankton. *L. mariae* occupy progressively deeper waters, reaching depths of 215 m on the Zambian shelf, where oxygen concentrations were as low as 0.6 p.p.m. *L. mariae* migrate into the surface waters to feed on clupeids at dawn and dusk, undertaking migrations of up to 200 m within an hour. *L. angustifrons* are solitary, sedentary and primarily benthic, although they are occasionally found aggregated in the pelagic zone, feeding on clupeids. Both species have also been found as adults, several kilometres up the Malagarasi river (De Vos and Snoeks, 1994).

Basic population parameters are summarized by Coulter (1991e, Table 6.IV). *Lates mariae* grow to around 75 cm, mature at 40–50 cm and live some 12 years. *L. angustifrons* reach over 1 m in length, live at least 12 years and mature at 45–57 cm. A few specimens weighing up to 80 kg and up to 2 m long have been caught, but these are now very rare.

Both species prey on a wide variety of benthic cichlids as adults (Coulter, 1991f, Table 8.XVII), but clupeids predominate in *L. mariae* stomachs during the period of maximum annual biomass of *Stolothrissa*.

Bagridae

Coulter (1991f) states that three genera of this family, *Chrysichthys*, *Bathybagrus* and *Auchenoglanis*, are represented in the benthic community. De Vos and Snoeks (1994) report a total of six genera from the lake, including the littoral habitat, the other three being *Bagrus* (*B. docmak*), *Phyllonemus* (three spp.) and *Lophiobagrus* (four spp.).

The six endemic *Chrysichthys* species form a species flock. Two are deep-water species: *C. stappersii* is common in the southern part of the lake, *C. grandis* is common on rocky bottoms. *C. graueri* and *C. brachynema* are found at depths of 20–80 m on the shelf areas, the former in rocky areas, the latter over a variety of substrates (Coulter, 1991f). Two other species, *C. sianenna* and *C. platycephalus*, are found in the littoral zone. *C. brachynema* and *C. grandis* grow to 20 and 5 kg, respectively, while the other species reach only around 1 kg. All species are omnivores, and all are capable of piscivory (although fish remains may come from dead, scavenged fish) but have variable diet compositions, with major dietary items including crabs, gastropods and shrimps in offshore species. Additionally, insect ostracods, bivalves, macrophytes and algae are found in the more inshore species (Coulter, 1991f, Table 8.XV, for review). Nothing is known of the reproductive ecology of these species, but they are likely to be deep-water spawners, as young have not been found in the littoral. The inshore *C. sianenna* may migrate into rivers to spawn (Bailey and Stewart, 1984). The eggs of all species are large and yolky, although ripe specimens are seldom found.

Bailey and Stewart (1984) also describe a catfish that resembles *Chrysichthys*: *Bathybagrus tetranema*, a monotypic genera. *B. tetranema* grows to 138 mm, has few, large eggs (a single 130-mm female contained 86 eggs of up to 3.5 mm diameter), rather like the mouthbrooding cichlids. Similar parental care may occur in this species, although its ecology remains unknown. It is found in waters 40–80 m deep (De Vos and Snoeks, 1994). *Auchenoglanis occidentalis*, a riverine catfish widely distributed in Africa, is found sporadically to depths of 140 m, but is more usually confined to near-shore areas in Lake Tanganyika. This species is an adaptable, generalized omnivore, and is one of the few such species to have penetrated beyond the littoral fringe in the lake.

Clariidae

There are seven clariid species in the Lake Tanganyika basin, of which five are riverine and widely distributed in tropical Africa, and two are endemic. *Tanganikallabes mortiauxi* is small, rare and is found in the littoral zone, while *Dinotopterus cunningtoni* is large, abundant and widely distributed in the benthic zone. The genus may be endemic and monotypic, although Greenwood (1961) has included the Lake Malawi *Bathyclarias* species flock (Jackson, 1959) in the genus, a revision not fully accepted among those working on these species (Roberts, 1975). Nothing is known about reproduction, but small juveniles are found in shallows close to the shore, indicating that reproduction probably occurs in coastal waters. Adults are widely distributed to depths of 120 m, but are most common in the sub-littoral. At night, *D. cunningtoni* migrates into the pelagic zone to feed on clupeids, and is sometimes caught in the pelagic fishery. The species also feeds on shrimps, but is caught by subsistence fishers in traps baited with weed (Poll, 1953), indicating omnivory.

Malapteruridae

Malapterurus electricus, which is widely distributed in lakes and rivers in tropical Africa, is not found in any of the Great Lakes except Lake Tanganyika. This bloated, blotchy catfish grows to 55 cm in the lake, and is commonest in the littoral, although it is found to a depth of 100 m. Its cryptic habitat and low catchability mean that it is probably more abundant than has been supposed. It feeds primarily on cichlids, and is a voracious and significant predator of the rocky sub-littoral fish community. Its reproductive biology is not known and there are conflicting reports as to whether it is a mouthbrooder (Coulter, 1991f), as it has large eggs. Larvae and juveniles of less than 3 cm have not been observed in the lake.

Taxonomic change may be expected, as *M. electricus* from elsewhere have been partitioned into a number of distinct taxa (T. R. Roberts and Norris, personal communication cited by De Vos and Snoeks, 1994).

Mochokidae

There are eight or nine mochokid catfish in the lake, all but one endemic (De Vos and Snoeks, 1994). Three are found mainly in the rocky littoral (*Synodontis eurystomus*, *S. petricola* and *S. dhonti*), while the non-endemic *S. nigromaculatus* is found mainly in rivers. *S. granulosus*, *S. lacustricolus* and *S. multipunctatus* are caught over a wide depth range, although they are most common in the 20–40 m band (Pearce, 1985b). The genus is characterized by the presence of serrated, locking dorsal and pectoral spines that are an effective deterrent to predators (Lowe-McConnell, 1987). In Lake Tanganyika, most of the genera appear to be molluscivores, with *S. multipunctatus* specializing in extracting the abundant gastropod *Neothauma tanganyicense* from its thick shell (Poll, 1953). *S. lacustricolus* feeds on both gastropods and lamellibranchs, while the diets of *S. nigromaculatus* and *S. granulosus* are not known.

Reproductive strategies seem to vary within the genera, with some species producing large numbers of eggs (over 3000 in *S. dhonti* and *S. nigromaculatus*), while *S. eurystomus* and *S. petricola* produce around 100 eggs. The most remarkable adaptation is shown by *S. multipunctatus*, which is a brood parasite of mouthbrooding cichlids (Sato, 1986). The eggs are incubated in the mouths of several host species together with the host's eggs, but hatch earlier. Having used up their yolk sacs, the catfish fry begin to feed on the fry of the host while still in the host's buccal cavity. Thus the early stages of development of this catfish not only depend upon their hosts for food and protection, but exploit almost their entire parental investment.

Cyprinodontidae

Lamprichthys is a monotypic endemic genus, and *L. tanganicanus* the only representative of the family inhabiting the lake; it is the world's largest cyprinodontid (De Vos and Snoeks, 1994). The species grows to around 150 mm in length and is found mainly inshore around rocks, but also ventures out into the pelagic zone where it is caught, together with the clupeids; it is an omnivorous planktivore (Poll, 1953). *L. tanganicanus* spawns above rocky areas, leaving its benthic eggs unguarded. The eggs are transparent, but still suffer high fish predation mortality (Brichard, 1989).

Aplocheilichthys, another cyprinodont, lives in lagoons and rivers in the lake basin.

Mastacembelildae

Twelve species of the sub-family Afromastacembelinae, one of three families of spiny eels, occur in Lake Tanganyika; eleven of these species are endemic. They therefore form the largest non-cichlid species flock in the lake. Most have cryptic, rocky habitats, and only one, *Aethiomastacembelus cunningtoni*, an omnivore reaching over 70 cm in length, is reasonably common in waters of at least 100 m depth (Pearce, 1985b), living on sand and mud bottoms as well as rock. *A. cunningtoni* is considered a fine food fish and is commonly caught on sandy bottoms in deeper waters (De Vos and Snoeks, 1994). Deep, rocky areas may well harbour some of the other species, but these areas have not been extensively explored. Congregations of large numbers have been reported, possibly spawning aggregations (Brichard, 1978; Pearce, 1985b). Abe (1989) records that *Afromastacembelus platysoma* males look after the eggs/juveniles in their rock burrows. Average brood size is six, although females produce at least 10 times this number of eggs. *Caecomastacembelus frenatus*, more abundant in the south of the lake than the north, is the only spiny eel also commonly found in affluent rivers (De Vos and Snoeks, 1994). *C. moorii* is more common in the north than the south.

The diet of seven of the 12 species of spiny eels has been studied from samples taken at Mbemba by Abe (1988b). The largest, *Caecomastacembelus moorii*, is piscivorous. The other species feed largely on benthic invertebrates, particularly shrimps and aquatic insects. Partial feeding niche segregation, both in diet compositions and in feeding mechanisms, was observed in the six benthivorous species.

The breeding season of the mastacembelids is thought to be November–February (Matthes, 1960).

Reproductive strategies of the offshore non-cichlid fish

Very little is known about the reproductive ecology of most of the non-cichlid fish in the deep benthic habitats of the lake. Coulter (1991f) places the fish into three categories based on the place where spawning is thought to occur. This sometimes differs from the normal habitat of the adult. The non-endemic species are likely to be riverine spawners, and some of the endemic *Synodontis* may have retained this spawning habitat. *Chelaethiops minutus* (Cyprinidae) and *Lamprichthys tanganicanus* (Cyprinodontidae) are known to come inshore to spawn in the littoral. As the only species spawning in the rocky littoral known not to afford parental care to their offspring, survival of the young depends on the transparency of the eggs, the ability of the young to feed exogenously on hatching, and tight schooling behaviour immediately on hatching (Brichard, 1978).

The bagrid catfish are thought to be deep-water benthic spawners. However, since there are no records of the two endemic species of clariids entering rivers it has to be assumed that these spawn in the littoral. The two benthic *Lates* species and *Afromastacembelus* species are also likely to be deep, benthic spawners, with the latter probably favouring rocky areas.

1.4.5 Littoral fish communities

The littoral habitat is more complex than the offshore benthic, with rocky and rocky/sandy shores providing a great variety of habitats. The littoral communities are dominated by cichlids, except in the region of river mouths where cyprinids and siluroids are more diverse. These groups also dominate the seasonal lagoons and swamps in river and stream deltas. There is a growing body of work on fish community structure and food webs in Lake Tanganyika, with particular reference to rocky littoral habitats.

A survey of 450 km of coastline in Tanzania was undertaken by a Belgian group in 1992 (Snoeks *et al.*, 1994). The survey concentrated on littoral fish communities at a depth of 0–5 m, and was designed to examine species distribution and taxonomy, including examination of geographical variation in morphological characters and genetic variability from mtDNA studies.

Recent Japanese work, largely on littoral, cichlid-dominated communities, is notable (e.g. Matsuda *et al.*, 1994). This research may provide baseline data, a source of comparison for inventory studies, and theoretical approaches to determining the likely effects of disturbances on fish communities.

The fish communities of sandy-bottom and sand/rock areas in the Mbemba region have been studied by Yuma (1989).

The rocky-littoral fish fauna off the Mahale mountains, south of Kigoma, have been surveyed (Takamura, 1983a). They are overwhelmingly dominated by cichlids, with over 87 species recorded. The next-most speciose group were the Bagridae, with six species. A full species list is given. Detailed studies on the depth and substrate-related distribution of fishes at Myako, in the Mahale Mountains National Park, are reported by Kuwamura (1987). More than half of the species are found in the top 5 m, with mouthbrooding cichlids dominating the rocky areas. Substrate spawners predominated in slightly deeper water, over sand/mud substrates. The Myako stream was also sampled, and seven fish species (six cyprinids, one cichlid) were reported. The food web for the littoral fish community at Myako is described by Hori (1987), who stresses the importance of mutualistic and commensal interactions in maintaining functional stability in the Lake Tanganyika fish communities.

Concurrent surveys of fish diversity in the north-western part of the lake are reported by Mihigo (1983). Surveys were carried out in five regions: Ruzizi river-mouth, Uvira, Kigongo, Luhanga and Pemba. A total of 177 species were recorded, many unique to particular stations. Luhanga had the highest diversity (89 species, 86 endemic), with a cichlid-dominated community. At the Ruzizi station

49 species were recorded, 35 of which were endemic; the fish community was dominated by non-cichlids, many of which were found in lagoon areas in the Ruzizi delta.

The results of surveys undertaken at Luganga and Ruzizi estuary in 1979–80 have been published more fully (Kawabata and Mihigo, 1982) and are a useful source of species lists. This survey confirms the findings of other papers that the rocky shore sites, dominated by cichlids, contain the most diversity (in species number) and the highest proportion of endemics. The Ruzizi site contains a number of riverine and specialist estuary/lagoon fish, as well as a number of lacustrine species, but overall diversity and endemism are lower. The exact sampling sites are not given in detail, so replication of these surveys and direct comparisons of surveys may be difficult to make.

The fish communities in the lagoons are dominated by cyprinids, particularly *Barbus* sp. (Kwetuenda, 1983). Nearly 70% of the recorded non-cichlid species and 41% of cichlid species recorded in the lagoons were thought to reproduce preferentially in this habitat during the rainy season, and large numbers of fry and juveniles were caught in samples.

Many essentially riverine fish also utilized the lagoons as breeding and nursery grounds. Surveys of three temporary lagoons over the period 1981–83 indicated that the species composition varied markedly, both between adjacent lagoons and between years (Kwetuenda, 1987), with the species present being determined by chance colonizations and fluctuations in environmental conditions, particularly in rainfall patterns.

A later survey (1987–88) added some further stations in the region of the Ubwari peninsula, where cichlids dominated the fauna at all stations, but six mastacembelid species were recorded (Sato *et al.*, 1988). Quite intense predatory activity by water cobra and otter was recorded in the region, and may have had an influence on the fish community structure.

A repeat of the 1979–80 survey of littoral fish communities at five stations in the north-west of the lake was carried out in 1990 (Nakaya, 1991): 109 species were recognized and a number of unidentified species recorded. Thirty-two species of non-cichlid were reported, representing 11 families. A full species list is given.

The littoral fish communities in the southern part of the lake, in Zambian waters, were sampled at 20 sites covering a variety of littoral habitats, at depths from 0–35 m, using SCUBA surveys, gill nets, hand-nets and anaesthetics (Kuwamura and Karino, 1991). Highest diversities were observed in the western part of the sample area (Mushi to Sumbu). As in the northern part of the lake, the cichlids are the most speciose, with a number of species not found in the north being located, as well as several undescribed or unidentifiable forms. Most of the non-cichlids were not identified to species, but eight families were represented.

The occurrence of a jellyfish (*Limnocnida tanganicae*) in the littoral zone at Kasenga, Zambia, was observed to cause the temporary disappearance of most of the littoral fish community, possibly because of de-oxygenation of the water column due to mass mortalities of the jellyfish (Karino and Kuwamura, 1989).

Vertical distribution of fishes on rocky shores has been studied at Kasenga (Karino, 1991), and the deep-water fish communities have been studied in the southern part of the lake as part of a repeat of the series of gill-net research programmes conducted every 10 years by the Zambian Fisheries Department (Phiri, 1991b). A total of 28–37 species were represented in catches from 20–100 m depth. These data provide useful comparisons with previous survey data in the same area, dating back to 1960 (see Coulter, 1991f for review).

None of the major affluent rivers has been systematically explored, although some sampling has taken place recently in the Ruzizi and Malagarasi systems (De Vos and Snoeks, 1994), and the Zambian Fisheries Department sampled the Lukuga River in the 1960s. The highest number of families of fish (at least 18) is represented in the rivers and marshes of the Lake Tanganyika basin (Coulter, 1991b, Table 10.I), although the number of species is lower than in the lake itself, and the proportion of endemics is also lower.

Non-cichlids are dominant in riverine, swamp, lagoon and estuary habitats. Recent sampling by De Vos and Snoeks (1994) showed that 103 of the 145 currently known non-cichlid species from the Tanganyika basin are found in affluent rivers or their associated swamps and marshes. The Malagarasi drainage harbours a particularly diverse non-cichlid fish fauna (70 species). The Ruzizi system, with 30 species, also contributes significantly to the diversity of fish species. The limited sampling that has taken place in small streams suggests fairly species-poor fish communities, but more extensive sampling may reveal a different picture.

Of the fish generally found in the littoral and associated habitats, and not reviewed in the above sections, there are: *Protopterus aethiopicus* (Lepidosierenidae), the only lungfish in the lake, found mainly in the Ruzizi delta and feeding on *Thalamus* and *Elaeis* palm-tree fruits (Brichard, 1989). The male builds a nest among reeds in lake-shore swamps and lagoons during the wet season, and attracts the female, which lays 2000–5000 eggs; both parents guard the young (Matthes, 1960).

Two species of *Polypterus* (Polypteridae) are found in the swamps of the lake basin: *P. endlicheri congicus* and *P. ornatipinnis*. The former is recorded from various swamps (Brichard, 1989), the latter from the Malagarasi swamps. Both are large predators, reaching 100 and 60 cm, respectively. The eggs are found attached to vegetation in fringing swamps and lagoons. The young, which stick to vegetation by means of an adhesive disc, spend the first dry season in lagoon habitats, isolated from the lake and associated rivers (Matthes, 1960).

The anabantid *Ctenopoma muriei* also spawns in swamps and lagoons. It does not nest and the eggs float freely among the plants; the young hatch 24 h after fertilization (Matthes, 1960).

The Kneriidae, which are confined to upland streams, are the subject of a recent taxonomic review by Seegers (1995).

Although several species of Mormyridae are found in the Zaire basin, only *Hippopotamyrus discorhynchus* has been recorded from the lake, at Cape Magara (Brichard, 1989).

Of the Characidae, two species of 'tigerfish', *Hydrocynus vittatus* and *H. goliath*, have been recorded from the lake. Neither is endemic, and they are confined largely to the vicinity of river mouths in the southern part of the lake. Both are large piscivores, reaching at least 10 kg in weight. The large *Alestes rhodopleura* and *A. macropthalmus* are sometimes seen in the lake, but are more common in rivers. *Bryconaethiops boulengeri* and *Micralestes stormsi* have been recorded from the mouth of the Lukuga outlet. The largest of the Citharinidae, *Distichodus sexfasciatus*, is also found around the mouth of the Lukuga. *Distichodus sexfasciatus* is distributed in the Zaire river basin, including Lake Tanganyika. Aquarium observations (Baute *et al.*, 1992) indicate that it is a daytime feeder, and is territorial.

Most of the Cyprinidae are riverine, and at least 36 species are represented in the lake basin. Several are found in the inshore waters of the lake. *Varichorinus* species are found all around the lake shore in the surf zone (Brichard, 1989). *Varinchorhinus* (now *Acapoeta*) *tanganicae* forms part of the aufwuchs-feeding community (Takamura, 1984) and specializes in feeding on filamentous algae, often from sandy substrates. Several *Barbus* species are found in the river mouths and around sandy shores. In the Myako stream, Muhale Mountains, the cyprinid *Raiamas moorei* was found in both lake and stream as adults, while *Acapoeta tanganicae* and *Barbus tropidolepis* were found in the stream as juveniles and in the lake as adults. Other species were confined to the stream (Kuwamura, 1987).

Representatives of the Amphiliidae, Schilbeidae and Tetraodontidae are also found in the Tanganyika basin (Coulter, 1991f).

1.4.6 Threats to diversity of non-cichlid fish

General processes

Threats to biodiversity by pollution, sedimentation, oil exploration activities and general habitat degradation are detailed in other parts of this study, and have been reviewed by Lowe-McConnell *et al.* (1992). Other than degradation of river habitats, fishing activities pose perhaps the greatest threat to the non-cichlid fishes.

Pollution

Many non-cichlid fish are found in rivers, or near river mouths. Land-use changes and human activities that affect river water quality are therefore a particular threat to many of the non-cichlid fish in the lake. Pollution threats include municipal waste from the Bujumbura area of Burundi (Sahiri, 1991) and sedimentation due to increasing deforestation, particularly in the northern part of the lake (Cohen *et al.*, 1993), making the Ruzizi River fish particularly vulnerable.

The concentrations of zinc, copper and selenium in *S. tanganicae*, *L. miodon*, *L. stappersi*, *Oreochromis niloticus* and *Boulengerochromis microlepis* from Burundi waters have been determined by atomic absorption spectrophotometry. In 1988, these concentrations were low (mostly less than 1% of World Health Organization permissible levels), indicating that the lake was not polluted with respect to these metals. Instead, fish were regarded as an important source of these essential trace metals in human diets.

Natural background levels of these and other elements (manganese, iron, lead, cadmium, mercury and arsenic) were established from samples of *S. tanganicae* and *L. stappersi.* Higher concentrations in the smaller *S. tanganicae* indicated that absorption of dissolved metal ions through the gills was a more important accumulatory mechanism than ingestions in food, or the larger, predatory *L. stappersi* would have higher concentrations of the metals in their tissues. Levels of all metals were low, indicating that the waters of Lake Tanganyika were unpolluted, although the date that the study was carried out is not given. An earlier study of organochlorine insecticide levels in fish species (Deelstra, 1977) also indicated low levels.

These studies provide a useful baseline from which to assess the extent to which heavy metal and organochlorine levels in the lake have altered over the last 10–20 years.

Fishing practices

Description of the fisheries

A description of the fisheries of Lake Tanganyika is given by Coulter (1991d). The existing fisheries are for food and for aquarium fish (see section 1.3.4); there is no significant sport or recreational fishery. Most of the fishers are subsistence farmers, and patterns of fishing activity are determined partially by the demands of farming activities. Fishing activities are centred around small, scattered villages or fishing camps, usually located where there is a sheltered beach or bay with cultivable land nearby; major centres of population are few (Coulter, 1991d). The impact of inshore fishing activities on fish communities is thus likely to be greater in areas with sandy, reed-bed or mixed sand/rock areas, than in the areas where the shoreline is open and rocky and the slopes are precipitous.

There are three broad categories of fishery:

- subsistence, often undertaken part-time by farmers to generate some cash or supplement diets;
- artisanal, full-time fishing, with the aim of generating income, undertaken as small-business ventures, distributing fish through small, local trading operations;
- industrial, capital-intensive mechanized fisheries, generally using purse-seiners and distributing fish through large markets, some international.

A variety of gears are used to catch the pelagic species (Figure 1.5).

Subsistence fishers operate close to shore, from small canoes, and they exploit mainly immature stocks of clupeids. Most fishing is done at night, with light attraction, using a conical-shaped scoop net, the *lusenga* or *agawe-shuro* net (Figure 1.5). *Lates* spp. are also attracted to lamps during fishing activities, probably by the aggregation of their clupeid prey (Coulter, 1991d). Subsistence fishers also operate handlines, basket traps, bottom-set gill nets and beach seines, all fishing in relatively shallow water near to shore. Of these, beach seines are the most important, and are employed extensively in the southern half of the Tanzania sector (Chapman and Bazigos, 1974) and in Zambia (Pearce, 1985a). Beach seining has largely replaced *lusenga* fishing, and the intensity of beach seine fishing along parts of the coast has caused concern about its effects on juvenile cichlid fish stocks (Roest, 1988) that regularly make up a large proportion of the catch (Ndaro, 1990, 1992). Other main catches are *Limnothrissa miodon* juveniles, and, more sporadically, *Stolothrissa tanganicae* juveniles.

Artisanal fishing uses canoe-catamarans that deploy lift nets to catch light-attracted schools of clupeids (Figure 1.5), or use light-attraction boats and open 'plank-boats', usually powered by outboard, to set *Chiromila* seines, a technique adapted from the fisheries of Lake Malawi. Once again, the target for these types of fisheries are the pelagic clupeids. The artisanal sector has been supported by development projects, with the aim of increasing catches and encouraging fishers to operate further from the shore, thereby catching larger clupeids than the inshore *lusenga* and beach seine fisheries (Coulter, 1991c). The most significant development has been the expansion of the improved 'Apollo' lift nets which are used widely on catamarans. They are now probably the single most important gear.

The industrial fishery uses 15-m purse-seiners adapted from those used in the Mediterranean by Greek fishermen. The seine is shot around a light-attraction boat. The fishery initially used large mesh sizes to target *Lates*, but now uses small meshes to catch a mixture of clupeids and *Lates*. Around 60 vessels were operating in the 1980s, but were confined largely to a few ports with access to markets (Bujumbura, Kigoma, Uvira) due to poor communications along the lake shore. Thus much of the lake is not exploited by industrial fishery.

Demersal and littoral fisheries

The demersal fisheries in the lake are insignificant when compared with the fisheries for clupeids and *Lates stappersi*, but are of local importance to subsistence fishers. This is in marked contrast to the fisheries of most African lakes, where bottom-set gill nets are often the most important fishery. Coulter (1991f) considers that "the scope for development [of a benthic gill-net fishery] appears limited, and the value of the fishery competes with the extraordinary scientific value of this fish community".

Nylon gill nets and engines to power canoes were introduced in the 1960s, and the number of gill nets operating in Zambian waters rose from 360 in 1964 to over 2000 in 1972 (Kendall, 1973b). The number of nets fished declined through the 1970s due to declining gill-net catches and increasing use of beach seining for clupeids. Much of the exploitation was in the south-east arm, as the south-west arm of the lake shore is within the Nsumbu Wildlife Park, where fishing is prohibited; differences in fish community structure between the two areas indicated that movement of fish populations between the two areas was limited (Coulter, 1991f). Comparison of gill-net catches in three surveys spanning 20 years reveals a now-familiar story in multispecies fisheries: a decline in catches of larger species, with some increase in numbers of small, prey fish failing to make up the loss of the larger species, both in total weight and in economic terms. Smaller mesh nets were being used to target the now smaller fish, but fishing moved further inshore, as the offshore areas were unprofitable. As a result, more juvenile *Lates* were being caught, perhaps precipitating the decline in adult stocks.

While it is sometimes supposed that driving a fishery towards exploitation of small fish feeding low in the food chain will increase yields (as may have occurred in the pelagic fishery with the removal of *Lates microlepis* and the two largely benthic species, resulting in higher clupeid catches; Coulter, 1991e) this has not been the case in the gill-net fishery of the south-east arm of the lake, as the populations of mouthbrooding cichlids are perhaps limited more by their reproductive capacity than by predation. Coulter (1991f) believes that the Zambian gill-net fishery has little potential for expansion in the south-east arm, and should be limited. In the lightly exploited deep benthic zone of the south-west arm, gill-net fishing should be restricted, together with industrial purse-seine fisheries that also take the large benthic piscivores and Bathybates (Cichlidae) that migrate into the pelagic zone at night to feed on clupeids.

There are few studies on the species composition of catches to inshore, artisanal and subsistence fishing activities. These fishing activities, although responsible for a small (but unknown) proportion of total fish landings from the lake, are likely to have greater impact on biodiversity conservation initiatives than the offshore, semi-industrial pelagic fisheries.

One of the few documented studies is that of Ndaro (1990, 1992) on the species composition of experimental beach seine catches in the Kigoma region (Tanzania). Gear deployed within the 20-m contour sampled 71 species belonging to 48 genera and 15 families. More than half the number of species caught were cichlids, mainly juveniles. Non-cichlid fishes accounted for between 3.5 and 75.2% of catches by weight (averaging over 20%), with the highest proportions found close to the Malagarasi River inflow. Most numerous species were *Auchenoglanis occidentalis*, *Hydrocyon lineatus*, *Citharinus gibbosus*, *Distichodus* sp., *Alestes macropthalmus* and *Synodontis multipunctatus*.

Fish densities were considered low, and attributed to intense beach seining activity directed at clupeids. These fisheries take place mainly on moonless nights with light attraction.

Catch statistics and the pelagic fishery

Fishery catch-effort statistics for the pelagic fisheries off Burundi for the period 1950–71 are given by Mann and Ngomirakiza (1973). The catches are broken down by traditional, artisanal and industrial fisheries, and by clupeids and predators; unfortunately stock assessment studies did not run in parallel with the efforts to develop and invest in the fishery. By 1970, signs of potential overfishing were detected, and regulations restricting fishing effort were recommended.

Total catches recorded from Lake Tanganyika between 1973 and 1986 were between 80 000 and 100 000 t/year. Table 7.1 of Coulter (1991d) provides a summary of catch data by fishery sector and country in years between 1973 and 1986 when catch statistics were considered reliable. The inshore subsistence fishery, based largely along the Tanzanian coast, took the majority of the catch during this period; around 10 000 canoes are engaged in the fishery. Artisanal fisheries have developed in northern Republic of Congo and in Zambia, and industrial fisheries were important in Zambia and Burundi, although more recently the industrial purse-seine fisheries in Burundi have been out-competed by artisanal lift-net fisheries (Petit and Kiyuku, 1995). Potential yields from the lake's fisheries, estimated by a variety of methods (reviewed by Coulter, 1991d, pp. 148–150) have been given as 380 000–460 000 t, but suggested conservatively as 330 000 t, largely from *Stolothrissa tanganicae* and *Lates stappersi*. These yields have not been realized, and it is not yet known to what extent these estimates may have been inflated by the assumption of relative homogeneity in the distribution of fish in the pelagic zone. Moreover, fishery statistics often record catches by broad categories, such as *ndakala*, that include both clupeid species and *L. stappersi* juveniles.

A review of fisheries in the north-west of the lake, centred in Uvira, indicated a change from dominance of industrial landings to artisanal fisheries during the early 1980s. Subsistence fishing around the lake shore and coastal lagoons accounted for only a small percentage of total landings, which were dominated by clupeids and *Lates* species (Enoki *et al.*, 1987). The Japanese/Republic of Congo team at Uvira made a number of studies of fishery catch/effort statistics and factors affecting catch rates in the artisanal fisheries (papers in Kawanabe and Nagoshi, 1987; Kawanabe and Kwetuenda, 1988). A review of the evolution of the pelagic fisheries in the south-east arm of the lake (Zambian waters) is given by Pearce (1995). The artisanal fishery in 1964 operated 650 *lusenga* nets with lights, from canoes. By 1990, the *lusenga* nets had been completely replaced by 195 beach seines, using 1436 lights. Pearce calculates that the effective artisanal fishing effort has tripled. Following a period of stagnation in the 1970s, the industrial fishery, which began in the 1950s, increased in the period 1980–91 from two to 17 purse-seiners. The area fished has increased from 300–1200 km^2, but total catches have been declining since 1985. The increased area fished is due to seasonal movement of the fleet following *Lates stappersi* stocks; clupeid stocks are still fished in a similar area.

A study of artisanal fishers was conducted in 1990 by Petit and Kiyuku (1995) in Burundi waters, northern Lake Tanganyika. The size of the artisanal fishing fleet, which uses lift-nets operated from canoe-catamarans, fluctuated between 700 and 1000 boats during the 1980s, with 750 being active in 1990; 70% of these had outboard motors. Piracy was a problem, with engined boats being a primary target, and the deterrent effect on people considering employment in fishing was thought to be a limiting factor in fishery expansion. The artisanal fishers are now targeting offshore stocks of *L. stappersi*. The increased efficiency of the artisanal fleet has had a detrimental effect on the industrial fishery sector, whose catches have decreased from 6000 t in 1980 to 2000 t in 1990, while the artisanal fishery has expanded gradually from around 12 000 t in 1980–81 to over 16 000 t in 1989–90.

Fishery assessments for Burundi waters are available for 1991–93 from Coenen and Nikomeze (1994). The situation in the industrial fishery is very different from that in the southern lake, with clupeids still being the predominant species in the catch (67–69%). Catch per unit effort has been decreasing since 1987.

An aerial frame survey in 1993 counted around 14 000 canoes, with more than half of those being in Republic of Congo (Hanek and Kotilainen, 1993). The same survey was used to provide a classification of the shoreline (Coenen, 1993). The most significant result is the apparently limited distribution of marsh habitats, which may provide important nursery grounds for juvenile *Lates*. The major areas are east of the Ruzizi, south of Kigoma, around the Malagarasi estuary and near Mpulungu.

Clupeid catches in the industrial fisheries off Kigoma declined before the recent decreases witnessed in Zambia and Burundi. The limited habitat for spawning around the steeply shelving shores, and lower productivity in the central lake, may adversely affect the reproductive capacity of clupeids, so that maintenance of a large adult biomass may be more important than in other regions, where variations in predators and food supply may affect recruitment to a much greater extent than parent stock biomass.

Total catches from the lake have recently been between 130 000 and 170 000 t (51 kg/ha/year) in 1992. Fishing effort, in terms of estimated number of fishing units, has remained largely unchanged since the 1970s; however, with both the subsistence and industrial sectors in decline, fleet compositions have changed while the artisanal sector continues to expand (Coenen, 1994). Frame surveys of fishing activities in the four riparian states were conducted in March 1995.

Fisheries and the pelagic ecosystem

The main ecological impact of the pelagic and offshore benthic fisheries has been to reduce the populations of the three larger *Lates* species to low levels, which has allowed the smaller *L. stappersi* to increase greatly in numbers in recent years (Roest, 1988; Coulter, 1991b). This appeared to occur in two phases in the Zambian and Burundi fisheries: firstly, the reduction in predators was followed by an increase in clupeid catches, then an increase in *L. stappersi* abundance led to a cyclical relationship between abundance of *S. tanganicae* and *L. stappersi* in pelagic purse-seine catches (Roest, 1988; Coulter, 1991b,c). This has been quantified for Burundi waters by Roest (1992) who demonstrates a significant log-linear regression relationship between the number of *Stolothrissa* recruits, the abundance of the parent stock, and the predator abundance.

It is normal in a size-unselective multispecies fishery that the larger species, with several age classes, will be fished out. *Lates* populations survive in areas that are not intensely fished: further inshore, or away from the centres of fishing activity, largely off Kigoma, the shelf area in the south-west of the lake, and the northern basin in Burundi waters. The general effect of the fishery has been to alter the pelagic fish community from a stable, predation-dominated system, to an unstable system where species abundance is determined primarily by competition for limited food resources (Coulter, 1991e). This may constitute an example of a shift from top-down to bottom-up control of the clupeid fish resources, but the fact that the altered system is maintained by the effects of another predator (the fishery) indicates that it is still likely to be controlled by top-down processes, with the apex predator now being the fishery.

The fishery in the southern part of the lake has had a dramatic effect on the pelagic fish community, well documented in Pearce's 1995 study. Before the fishery developed, the catch was composed of 50% clupeids and 50% *Lates* (Coulter, 1970). The fishery has since been through three stages, during which the number of species on which it depends has been reduced from six to one. Firstly, there was a reduction of large *Lates* species; these were replaced by *Lates stappersi* (1962–66). Secondly came a phase when the fishery was based largely on *S. tanganicae* and *L. stappersi*, *L. miodon* having been reduced in importance due to overfishing by artisanal fishers. During the 1970s, there was an inverse cyclic relationship between catches of *S. tanganicae* and *L. stappersi*. Thirdly, a reduction in clupeids in the fishery occurred: since 1986 62–94% of the total pelagic catch has been *Lates stappersi*. This switch to a single-species fishery has been accompanied by a recent decline in total fish biomass. Pearce (1995) suggests a strategy for restoration of the fishery:

- reduce inshore fishing pressure on juvenile *L. miodon* – this species is more common inshore at all stages in its life cycle and has suffered tremendous fishing pressure by artisanal beach seining;

- maintain high fishing pressure on *Lates stappersi*, while avoiding exploitation of *S. tanganicae* – as the two species are often caught at different times of year, this is a matter of adopting a seasonal fishery until clupeid stocks are rebuilt.

The aquarium trade

There are few non-cichlid fish from Lake Tanganyika that are of interest to the aquarium trade. The following information on those that are popular or potentially popular aquarium fish is taken from Brichard (1989).

Distichodus sexfasciatus (Citharinidae), a distinctive orange and black-banded fish found around the mouth of the Lukuga outlet, is a sought-after aquarium fish, although it is fast-growing, reaching 1.5 m in length and 25 kg. Several other, smaller species are found in the rivers of the lake basin. Some of the small, inshore cyprinids (*Barbus*, *Barilius*) are also of some interest to aquarists.

Two small bagrid catfish, *Phyllonemus typus* and *P. filinemus*, found on rocky shores, are considered important additions to tanks of Tanganyika cichlids, due to their scavenging role as well as their attractive appearance. They do not breed in aquaria and little is known of their biology. *Chrysichthys sianenna*, the smallest member of this endemic bagrid species flock, is also collected for aquaria. All the mochokiid catfishes are popular aquarium fish, with *Synodontis multipunctatus* and *S. granulosus* being particularly attractive. The latter species is very rare and fetches high prices.

The small cyprinodontid, *Aplocheilichtys pumilus*, found in fringing lagoons and swamps all around the lake shore, is a hardy, active fish, easy to spawn and popular with novice aquarists. *Lamprichthys tanganicanus* is delicate and difficult to keep, but is described as 'incredibly beautiful'.

Of the endemic species flock of spiny eels (Mastacembelidae), *Mastacembelus* (*Aeithiomastacembelus*?) *tanganicae* is the most popular with aquarists. *A. platysoma* is also sought after, but is very rare.

The impact of the aquarium trade on fish communities has not been documented, but if trade statistics are made available, they will provide data on species and area-specific catches that will be of tremendous value.

Introduced species

Both exotic tilapiine cichlids and Asian aquarium fish are known to be established in Lake Tanganyika (G. Ntakimazi and L. De Vos, unpublished data, cited by Cohen, 1994). The extent to which they occur and the possibility of limiting their expansion should be considered. Since 1988, a small and very prolific species of tilapia, *Oreochromis leucostictus*, has appeared in the Ruzizi estuary and associated swamps (Lowe-McConnell *et al.*, 1992). There is a fear that carp (*Cyprinus carpio*) introduced into the Ruzizi drainage rice fields could enter the lake.

1.5 CRUSTACEAE

1.5.1 Copepoda

Taxonomy

The Copepoda are split into four orders which are all represented in Lake Tanganyika:

- Calanoid one species which is not endemic
- Cyclopoida 47 species of which 17 are endemic
- Poecilostomatoida 4 species of which 2 are endemic
- Harpacticoida 16 species of which 14 are endemic.

At present 68 species of copepoda are known of which 33 are considered to be endemic. According to Dumont (1994a) the only calanoid species, *Tropodiaptomus simplex*, is endemic to Lake Tanganyika. Extensive research has been done in the pelagic habitat and to a lesser extent in the inshore habitat of Lake Tanganyika and new species are not expected. However, other water bodies in Central Africa have not been very well investigated and a number of species, now regarded as endemic in Lake Tanganyika may appear yet be recorded as present elsewhere.

Distribution

Geographical pattern

Only one calanoid species, *Tropodiaptomus simplex*, and six cyclopoid species are present in the pelagic waters. Most cyclopoid species are found in the inshore waters. The endemism of so many copepoda is remarkable as most African copepoda species are widely distributed over the continent. Many copepoid species in inshore waters are bottom dwellers, e.g. harpacticoids.

Abundance

The calanoid *Tropodiaptomus simplex* is the most abundant species of the zooplankton in the pelagic habitat of Lake Tanganyika and dominates both in number and biomass. Densities of this species vary with place and time and there are indications that a cycle in abundance over several years takes place. The abundance in number of zooplankton in the inshore habitat is not known.

Ecology

Predation

In the pelagic habitat the two clupeid species, *Limnothrissa miodon* and *Stolothrissa tanganicae*, are the main predators of the copepoda stock. In the inshore waters, a large number of cichlids depends on zooplankton, which in this habitat is dominated by cyclopoid species. The juvenile stage of the clupeid fishes which stay in the inshore waters depends heavily on this inshore zooplankton. Any change in the abundance of the zooplankton in the inshore areas can strongly influence the pelagic fish stock.

Ongoing research

The present FAO/FINNIDA project has as one of its priorities the study of the population dynamics of the zooplankton of the lake. However the results of this report are in progress and are not yet available. The outcome of the results of this research group will guide the next steps in research. The Japanese team from the University of Kyoto has carried out a number of case studies at several places in the inshore area of the lake as part of its ongoing ecological research. These have not yet been published. It may be noted that almost all zooplankton studies in Lake Tanganyika have concentrated on the pelagic habitat, while hardly any attention has been given to the inshore areas of the lake.

1.5.2 Branchiopoda – Cladocera

General

In the Lake Tanganyika basin, 24 species of cladocera are found. However, their paucity in number and species in the main lake, as well as the absence of endemics in Lake Tanganyika, is remarkable.

Distribution

All species of cladocera found in the Lake Tanganyika basin have a wide distribution in Africa, and all were found in the inshore area and the adjacent waters of the lake. Not one species was found in the pelagic habitat (Dumont, 1994b, c).

There is no need to search for cladocera in the pelagic habitat, as this has been done extensively since G.O. Sars noticed the absence of cladocera from Lake Tanganyika in 1909 (Coulter, 1991a). The distribution and abundance of cladocera in the inshore habitat, however, are not known.

1.5.3 Ostracoda

General

The ostracodes in Lake Tanganyika are important not only because of the high endemism, but also because changes in the past can be investigated by means of their valves, which are found in cores of the lake sediment.

Taxonomy

Existing data

There are 85 species of ostracoda described from Lake Tanganyika, of which 74 are regarded as endemic. However, there are in stock over a hundred undescribed species waiting for description or recognition as new species for the lake (K. Martens, personal communication). The number of species and genera for Lake Tanganyika is estimated at 200 and 25, respectively (Martens, 1994). A few species have been described recently (Wouters and Martens, 1992, 1994). These species were found during expeditions to the Tanzanian and Zambian coasts of the lake. The number of described ostracod genera is also still increasing (Martens, 1985). On the other hand, it is quite possible that species now regarded as endemic to Lake Tanganyika will, on investigation, also be found in other water bodies in Africa. This has been the case with, for example, *Tanganyikacythere burtonensis*, which was regarded as endemic to Lake Tanganyika but was later found in Lake Albert (Wouters and Martens, 1994).

Ongoing research

The Royal Belgian Institute for Natural Research, Brussels, under the guidance of K. Martens and K. Wouters, is the leading institute carrying out research into the taxonomy of ostracoda from Lake Tanganyika. Sampling in the lake has been carried out in combination with the Research Unit of Fishes from the same institute and the Mollusc Group from the University of Arizona, USA.

Distribution

Geographical patterns

Information on the distribution of ostracodes in Lake Tanganyika is scattered due to a limited number of sampling sites and a still-developing taxonomic knowledge. Some species have a distribution of the type locality only. For example, *Tanganyikacythera caljoni* is known only from Karona (Burundi) where it was sampled in a bamboo stand on sand at 1 m deep water, and *Cyprideis rumongensis* from the type locality in the delta of the river Dama near Rumonge (Burundi). Most species are collected in the littoral zone of the lake but there is evidence of their presence at deeper waters, e.g. *Cyprideis rumongensis* was found at 50 m deep.

Abundance

The abundance of ostracod species in Lake Tanganyika is not well known. From fish gut contents it is known that some taxa may be very abundant. Marked differences have been found in the number of species between habitats which are strongly, moderately or not affected by siltation (Cohen, 1994).

Habitat

A number of species are free swimming, for example, *Allocypria* and *Mecynocypria* genera, but most are confined to benthic habitats.

Ecology

Reproduction

All Tanganyikan ostracod species of the Cytheroidea are brooders, even those that have an externally visible brood pouch for eggs and larvae (Martens, 1994). Ten to eleven eggs were found in the pouches of two females of *Cyprideis mastai*.

Predation

Ostracodes are preyed upon by a number of cichlid fishes. Coulter (1991a) reports ostracodes as important food for deep-water-dwelling cichlids, e.g. *Xenotilapia nigrolabiata* and *X. caudafasciata*. The non-cichlid *Chrysichthys stappersi* were recorded with stomachs full of ostracodes.

1.5.4 Decapoda – shrimps

General

The shrimps of Lake Tanganyika have been neglected by most researchers and their role in the ecosystem has been poorly investigated. At present, increasing attention is being applied to this group and a number of activities are going on. Only part of the results have been published. Two groups of researchers are carrying out work on shrimps: the FAO/FINNIDA project is researching shrimps as part of the zooplankton of the lake; and a Japanese team from the University of Kyoto has carried out some investigations on shrimps as part of their ecological research in the Lake Tanganyika project.

Taxonomy

Available data

In Lake Tanganyika 14 species of shrimps are found. Most of the taxonomic work on these species was done a considerable time ago. It is remarkable that in one shrimp family, the Atyidae, three endemic genera have developed of which one genus, *Limnocaridina*, gave rise to eight species. In the only other Lake Tanganyika shrimp family, Palaemonidae, with one widely distributed genus *Macrobrachium*, only one endemic species is found.

Several colour patterns within one species are found among living specimens from different habitat types, which reflect the dominating colour of that habitat. These colour patterns seem to be effective in avoiding predation (Kimbadi, 1995).

Field identification of Lake Tanganyika shrimps is problematic as there are often only minute differences in characteristics. However, experienced field observers can distinguish some taxa on body shape and behaviour (Narita, 1995a).

Distribution

Geographical patterns

All species of shrimps from Lake Tanganyika can be found inshore, while only three species are found offshore: *Limnocaridina tanganyikae*, *L. parvula* and *Macrobrachium* (*Palaemon*) *moorei*. *M. moorei* is pelagic in its larval stage, moves inshore in its juvenile stage and is deep benthic in its adult stage (Pearce, 1991). The pelagic stock of shrimps is found lake-wide. The only non-endemic species, *Caridina nilotica*, is found in all waters bordering the lake, but not in the main lake itself.

Habitat

Atyid shrimps are the most abundant shrimps on the bottom substrate down to a depth of 25 m (Narita, 1995a,b). In the littoral habitat of the Zambian sector of the lake, *Limnocaridina latipes* was found to be the most common species (Narita, 1995a). *Macrobrachium moorei* was only found below a depth of 25 m (Narita, 1987). Most shrimps of the inshore zone are active at night. During the day most shrimps hide in rocky crevices and between algae. The pelagic stock of shrimps follows the diurnal vertical movement of the zooplankton.

Narita (1995b) found that shrimp numbers increased in the territories of mouthbrooding cichlids while these were actually brooding (i.e. not feeding or defending their territory against all cichlid intruders). The shrimps hide in filamentous algae.

Differences in length-frequency distribution of *Limnocaridina tanganyikae* between the pelagic and the littoral stocks have been found (Mashiko *et al.*, 1991).

Abundance

The most abundant species of the open water is *Limnocaridina tanganyikae*, which is observed in enormous swarms (Brichard, 1987). Pearce (1991b) indicates *Macrobrachium (Palaemon) moorei* to be the most important species in the Zambian part of the lake. Between May and October, but especially from June–August, local fishermen sometimes catch tons of this species in their beach seine-nets while targeting for clupeidae. In zooplankton studies larvae of the small pelagic prawn, *L. parvula*, formed 45% of the zooplankton biomass. As the southern part of the lake is regarded as being more productive than other parts, the stock size and the availability of shrimps may be more profound and its role in the ecosystem more important.

Ecology

Reproduction

All lacustrine shrimps from Lake Tanganyika have relatively small-sized eggs, which are kept for some time in a brood pouch between the pleopods of the female. The average number of eggs for *L. tanganyikae* was found to vary from 23–184 (Kimbadi, 1989) and 25–310 (Kimbadi, 1991). The mysis-stage larvae of the genus *Limnocaridina* are extremely abundant in the inshore waters.

There is no information on spawning periods or on seasonal spawning. However, as there is a period of mass abundance in the adult stage, it may be expected to be a seasonal spawner.

Food

Most shrimps in the lake are predators of zooplankton. The only exception to this is *Caridina nilotica*, which feeds on detritus.

Predation

Shrimps play an important role in the diet of the juvenile *Lates* species, while *Lates stappersi* may feed on shrimps throughout its life (Coulter, 1991a). H. Matthes (cited by Pearce, 1991a) found *Lates mariae* and *L. microlepis* of 1.5–6 cm feed exclusively on shrimps. Shrimps form the principal food for both clupeid fish species from 4 cm onwards (Phiri, 1991a). The whole littoral cichlid fish community feeds opportunistically on shrimps (Narita, 1987), while several specialist predators are known, e.g. *Lamprologus compressiceps* (Hori, 1987). The atyid shrimps constitute a food item in 20% of the fishes of the littoral habitat (Hori *et al.*, 1993). Shrimps are also an important food item in the deep benthic fish community (Coulter, 1991a).

It is noted that the role of shrimps in the food web of the southern part appears to be different from that in the northern part of the lake. The role of shrimps in the ecology of the lake generally underestimated. Stock size of the shrimps of Lake Tanganyika is not known.

1.5.5 Decapoda – crabs

General

Interest in the crabs of Lake Tanganyika has focused largely on their peculiar appearance. A team from the University of Arizona, under the guidance of Dr K.West, is conducting ongoing research on the Lake Tanganyika crabs in relation to the gastropod fauna of the lake (West *et al.*, 1991; West and Cohen, 1994).

Taxonomy

In Lake Tanganyika 10 species of crab are found. The status of two species is doubtful, as one is described on the basis of only one specimen, while a second 'species' is possibly a juvenile stage. All described species of crab from Lake Tanganyika belong to only one genus, *Potamonautes*, and all are relatively small-sized (less than 50 mm carapace length) (Coulter, 1991a). The Lake Tanganyika crabs differ from all other African fresh water crabs in having heavily calcified chelae.

Distribution

Habitat

Two species, *P. lirrangensis* and *P. loveridgei*, are found in the pools and rivers connected to the main lake. Five lacustrine crabs are found in the littoral and sublittoral habitat, where most species are to be found in burrows in the sand and mud between the rocks and under stones. Old gastropod shells are also used as cover.

Abundance

Brichard (1989) reports crabs to be common but does not mention any particular species. It was noted that most crabs are found in the zone at depths of 2–5.6 m, where densities of 15–17 animals per m^2 are common (Abe, 1988a).

Ecology

Reproduction

Females carry eggs and early developing stages in a pouch on the abdomen. There is no free larval stage. There are no data on periodic spawning, growth or size at maturity.

The number of eggs found in buried females varied from 167–1073 eggs in *P. armata*, and from 10–21 and 38–74 eggs in two unidentified species. The eggs remain attached to the female and even juveniles are carried for some time (Abe, 1988a).

Food

Lake Tanganyika crabs are specialized snail eaters. The impact of crab predation on the abundance of the gastropod species is still unknown.

Predation

Crabs are eaten by fish including *Chrysichthys* spp. and some cichlid fishes, e.g. *Lamprologus lemairii*. The impact of predation on the abundance of crabs is not properly understood.

1.6 MOLLUSCS

1.6.1 Taxonomy

In Lake Tanganyika 75 species of molluscs are found: 60 gastropod species of which 37 are endemic, and 15 bivalve species of which nine are endemic. The highest level of endemism is found amongst the true lacustrine species, while those species found in lagoons, pools, rivers, etc., are usually widely distributed outside Lake Tanganyika. The description of these taxa is mainly based on the morphology of the shell only. The present number of 75 species is based on the work of Leloup (1953) who had extensive collections at his disposal following the Belgian Hydrobiological Exploration of 1946–47. Leloup revised the large number of species (several hundred), of which J.R. Bourguignat, the first gastropod taxonomist to work on the lake, had already described 242 species.

Increased knowledge and improved techniques indicate the existence of more species than previously recognized. For example, a new genus and species of gastropod was found in 1995 in Kala Bay (Verheyen, 1995). The reasons for suggesting the existence of more species in Lake Tanganyika are, according to Michel *et al.* (1992), as follows:

- morphospecies differ in terms of reproduction and habitat preference
- some morphospecies coexist in the same habitat
- some groups have different internal anatomy, for example, different radula
- there are non-interbreeding groups, as indicated by electrophoretic data.

A number of the gastropod molluscs of Lake Tanganyika have remarkably strong calcified shells, and are described as 'thalassoid', in reference to the marine-like appearance of their shells. These thalassoid shells are thought to be the result of coevolution between gastropod snails and predating crabs (West and Cohen, 1991).

There is no up-to-date field guide with pictures and identification keys available for identification purposes.

1.6.2 Abundance and distribution

The distribution of gastropod snails in Lake Tanganyika was until recently not well investigated. Distribution patterns for a number of species are known from recent research done in Burundi, Tanzania and Zambia. However, almost the entire stretch of the Zairian coast has yet to be investigated. Horizontal distribution is determined by the type of bottom substrate and the degree of exposure to water turbulence in the wave zone. Vertical distribution indicates that most species are to be found in the littoral zone, while some species of the genus *Burnupia* have a range up to a depth of 70 m (Coulter, 1991a). The deep benthic fauna consists of a few species with thin shells.

Several mollusc species in Lake Tanganyika are common while others are extremely rare, and only very occasionally a living specimen of a particular species is found. The most common species is *Lavigeria paucisostata* (Yuma and Nakai, 1987). A good example of a rare species is demonstrated by the discovery of a living specimen of *Bathynellia* in Cameron Bay (Zambia) in 1995. The only previously recorded living specimen of this species was collected in 1890 (Verheyen, 1995). With regard to other species, the whole stock of a single locality morph may consist of a few hundred specimens (Cohen, 1991a).

Most species and morphospecies are found on the rocks in the littoral zone and their distribution may be limited by physical obstacles such as a river sedimentation area, for example, the mouths of the Ruzizi and Malagarasi Rivers as demonstrated by Michel *et al.* (1992). Unlike the cichlid fish community, there are no indications that historical fragmentation of the lake into two or three sub-basins played an important role in the diversification process of gastropod species.

Most gastropods in Lake Tanganyika are nocturnal feeders, while by day most snails can be found resting between cracks in the rocks.

1.6.3 Ecology

Reproduction

Information on reproduction is known from only some gastropod species. All species and morphospecies of, for example, the genus *Lavageria,* are brooders, while other species, such as those of the genus *Spekia*, deposit their eggs on rocks. A general characteristic of many gastropods of Lake Tanganyika is that they have a small number of juveniles. The number of eggs and broods found in the brood pouches of six species of *Lavigeria* varied from 21–68 for *L. grandis* to 163–315 for *L. nassa* (Yuma and Nakai, 1987). The female of the unionid mussel *Grandidieria burtoni* incubates her eggs in marsupia within the inner gills, and egg numbers can range from 52–624, depending on shell size (Kondo, 1987). Another mussel, *Moncetia lavigeriana*, was found to have eggs varying in number from 3000 to 22 000 (Kondo, 1985). This latter species has larvae which become ectoparasitic on fish following their release.

Predation

The molluscs of Lake Tanganyika are eaten by crabs and molluscivorous fish. The most important predators are crabs (West and Cohen, 1991). There are some specialized cichlids with molariform pharyngeal teeth, e.g. *Neolamprologus tretocephalus*, which prey upon juvenile snails.

1.6.4 Threats to the biodiversity of molluscs in Lake Tanganyika

The molluscan fauna of Lake Tanganyika is not exploited by man, so the question of over-exploitation does not arise. Apart from a general threat of habitat pollution by human and industrial waste, a severe threat comes from excess sedimentation of the littoral habitat as a result of topsoil erosion in the Lake Tanganyika watershed. This sediment first fills the cracks between the rocks and then successively covers the entire rocky habitat. The sediment in suspension reduces light penetration so algal growth is reduced and those molluscan species which depend on the rocky habitat disappear.

1.7 OTHER GROUPS

1.7.1 Cnidaria

Taxonomy

Available data

There are two cnidarian species in the Lake Tanganyika basin: the sessile, non-medusa-forming *Hydra vulgaris*, and *Limnocnida tanganyicae* which has a sessile stage and a free-living medusa stage. There is confusion concerning the number of cnidarian species on the African continent. However, this is of minor importance for the Lake Tanganyika species, as there is no doubt about the identity of *L. tanganyicae* which is described as originating from this lake.

Distribution

Geographical patterns

Limnocnida tanganyicae is found in several African water bodies but never in such great numbers as in Lake Tanganyika, where the species is found throughout the lake. *Hydra vulgaris* is a cosmopolitan species found in the waters adjacent to Lake Tanganyika but not in the main lake itself.

Habitat

During its free-living medusa stage, *Limnocnida tanganyicae* is found at inshore and offshore sites over the entire lake. During its sessile, hydroid stage it lives in sheltered waters attached to the stems of water plants such as *Phragmites* (Coulter, 1991a).

Abundance

In Lake Tanganyika, *Limnocnida tanganyicae* medusae are very irregular in abundance and vary from complete absence to sudden appearance in large swarms. An entire swarm may migrate to deep waters or to the surface, and rise or sink vertically in the water column at dusk and dawn. The appearance of large numbers of medusae is related to low densities of clupeidae. A whole inshore fish community at a rocky habitat can disappear into deeper waters when medusae appear in great numbers, and return when they leave (Karino and Kuwamura, 1989). The density of jellyfish can be approximately 10 individuals per litre of water in shallow waters less than 3–4 m deep.

Ecology

Reproduction

Sexual reproduction of the medusa begins in May and lasts until June. Ripe medusae are found until December (Bouillon, 1954 cited by Coulter, 1991a). Asexual reproduction of the medusa takes place throughout the year by budding, and the hydroid stage also reproduces asexually by budding of the medusa from the hydroid.

Food

As with all Cnidaria, the medusa of *Limnocnida tanganyicae* is an opportunistic predator. In Lake Tanganyika the eggs of the pelagic clupeidae may be their major food source (Dumont, 1994b,c). Zooplankton is probably of minor importance, as the only available species, *Tropodiaptomus simplex*, is fast-swimming and can avoid the hunting jellyfish which have a characteristic down-parachuting, hunting behaviour (Dumont, 1994b).

Predation

Exact statistics on predation of *Limnocnida tanganyicae*, as well as other mortality factors, are unknown.

1.7.2 Porifera

Taxonomy

Available data

There are nine species of sponges known from the Lake Tanganyika basin. Sponges in the lake, like many lower animal groups, have not yet been well investigated and possibly more species may be discovered.

Distribution

Geographical patterns

Of the nine sponge species, seven are endemic to the lake, while two non-endemic and widely distributed species have been described from the poorly investigated basin of the inflowing Malagarasi river. These two species are not found in the lake itself (Coulter, 1991a).

Habitat and abundance

All lacustrine sponge species are found attached to rocks and/or mollusc shells in the littoral zone, while some species are also found in the deep-water habitat. Brichard (1989) states that "sponges grow everywhere in the lake", and indicates that they are more abundant in the northern part.

Ecology

Little is known of the ecology of sponges from Lake Tanganyika. There are no data on growth, distribution preference, etc.

Predation

Predation of sponges by fish was never observed by Brichard (1989) nor were predation scars noted. Hori (1987) indicates that four cichlid species, *Julidochromis regani*, *Telmatochromis dhonti*, *Chalinochromis brichardi* and *Petrochromis famula*, and one catfish, *Synodontis eurystomus*, feed on sponges. Over 60% of the diet of *Julidochromis regani* is believed to consist of sponges.

1.7.3 Insects

Surveys of insects dependent upon the lake for at least part of their life histories are known to be incomplete. Several endemics exist, including a pelagic trichopteran, *Limnocepis tanganicae*, which is commonly attracted to the fishing lights on the lake, and four endemic Heteroptera which live under stones in shallow water (Coulter, 1991a). Some groups, such as the Trichoptera, have received considerable attention (Marlier, 1962), whilst most have not.

1.7.4 Plants

Phytoplankton

Phytoplankton collections from the lake have been made since the 1904–5 expeditions to Tanganyika (West, 1907). Based on subsequent work, 474 infragenic taxa of diatoms, 224 taxa of Chlorophyta, 111 taxa of Cyanophyta, 59 of Euglenophyta, 21 of Chrysophyta, 19 of Dinophyta, 14 of Cryptophyta, four of Xanthophyta and one taxon of Prymnesiophyta have been recorded from pelagic collections. Of the diatom flora of Lake Tanganyika, 8% are endemic, a high figure compared to algal situations in other aquatic habitats.

Epiphyton or Aufwuchs

Communities of algae are to be found on hard surfaces, particularly rocks, within the euphotic zone. They play a particularly significant role in the feeding niches of groups of cichlid fish species, and probably also of gastropod snails. These communities consist particularly of desmids, pennate diatoms and some blue-greens. As a group they appear to be more species rich than the algae of the phytoplankton but, as with the latter, endemism is very low (Coulter, 1991a). Given their important role in the littoral food chains of the lake, the epiphyton have received little attention.

Higher plants

Communities of leafy vegetation are comparatively sparse in Lake Tanganyika, unlike Lake Victoria, for example, and studies on plants have largely been lacking.

Most heavily vegetated areas are confined to the mouths of the rivers which, as in the case of the Malagarasi, may give rise to a large swamp. In these areas, the densest part of the swamp is typically dominated by *Cyperus papyrus*, *Typha*, *Carex* and other emergent plants. In centres such as the Malagarasi, water beneath these papyrus beds may be completely deoxygenated. The lagoons on the margins of these beds or in other marginal areas may support areas of abundant water weeds such as *Potamogeton*, *Ceratophyllum*, *Chara* and *Utricularia* and in places floating mats of water lilies (*Nymphaea*), water chestnut (*Trapa*), water fern (*Azolla*) and the Nile cabbage (*Pistia*) (Van Meel, 1952; Beadle, 1981).

The influence of vegetated areas may be disproportionate to their restricted distribution through their provision of cover for juvenile fishes and invertebrates.

1.8 GENERAL ISSUES FOR BIODIVERSITY

1.8.1 Measuring diversity

In trying to get initial estimates of biodiversity from rapid surveys, it has been suggested that taxonomic levels above the species be used as the basic unit of measurement (Gaston and Williams, 1993). This has advantages where species identification is difficult, or where many undescribed species are present. A study of diversity in Australian ants (Anderson, 1995) has indicated, however, that diversity of genera is a poor measure of species diversity in taxa where a small number of genera contribute a large proportion of the total number of species present. This would be the case in Lake Tanganyika, where many of the cichlid genera contain a large number of species, so the use of higher taxon categories cannot be recommended.

In the case of modelling studies to determine energy flow between major ecosystem components and to investigate the potential effects of pollution and fishing on system integrity, stability and productivity, aggregating taxa into functional groups is probably essential for the models to be tractable. This approach may also be desirable theoretically, as it is doubtful if the chaotic dynamics of individual species allow their abundance to be predicted, whereas the behaviour of guilds or functional groups is thought to be more predictable (Hay, 1994). This is confirmed by the apparent stability (resilience and persistence) observed for feeding guilds of littoral cichlid fishes in Lake Tanganyika (Hori *et al.*, 1993).

1.8.2 Modelling studies

Carbon sources supporting the secondary production in the food chain that forms the basis for pelagic fisheries are a focus of part of the Lake Tanganyika Research (LTR) programme (Salonen and Sarvala, 1994). Energy flows within the pelagic ecosystem of the Burundi sector of the lake have been quantified using a steady-state trophic model (ECOPATH) for the periods 1974–75 and 1983–84 (Moreau and Nyakageni, 1988; Moreau *et al.*, 1991). These approaches can be useful in determining the likely effects of change on various ecosystem properties (ascendancy, flow, connectance, maturity, ecotrophic efficiency, etc.). Another simple approach to monitoring aquatic ecosystem health is through use of biomass size spectra (Sprules and Munawar, 1986); a recent application to Lake Malawi and a review of size-spectrum theory are given by Allison (1996).

1.8.3 Protecting diversity

The establishment of reserves provides a useful solution to the problems associated with protecting biodiversity. Within reserve activities that are perceived as threats to diversity – fishing, deforestation, use of agrochemicals, human settlement – can be controlled. In selecting reserves, a common goal is to maximize the species diversity that can be protected. A central debate in the use of reserves for biodiversity conservation is as to whether a single large or several small sites are better for ensuring diversity conservation. The arguments have drawn on equilibrium theory of island biogeography and metapopulation dynamics to advance their respective cases (reviewed by Saetersdal *et al.*, 1993; Lomolino, 1994). In practice, the nature of threats to diversity differs, as do the life histories and distributions of the taxa present in different ecosystems, so that reserve selection is best based on extensive empirical data for particular regions, rather than generic prescriptions (Lomolino, 1994). Saetersdal *et al.* (1993) present three interactive methods for reserve selection based on empirical survey data on species distributions. Two of the algorithms developed were designed to maximize the diversity of species protected (MAXDIV, SPECMAX); a third (ENDEMIC) produces a priority list of conservation sites that maximizes the representation of particular groups of species, e.g. endemics. In the example given for a Norwegian deciduous woodland, preserving 10% of the area would conserve 50% of regionally rare species, while preserving 50% of the area would conserve 90% of rare species. This type of program (available from the authors as FORTRAN programs) can be used to aid decision making processes, but will have to be supplemented by inclusion of studies on the logistical feasibility of preserving particular areas, given prevailing socio-political conditions.

It can be argued that the focus on reserves is misplaced (Witting and Loeschke, 1995), as it concentrates available resources and protection on areas which contain large populations of common species, and small populations of rare ones whose continued existence may be threatened by their vulnerability outside reserve areas. It is essentially a question of whether the reserves are sufficient to protect a minimum viable population, with attendant difficulties in identifying this population size. Conservation efforts should therefore be directed at minimizing future loss of biodiversity from an overall area, in this case, Lake Tanganyika and its catchment.

The threats to biodiversity that have been identified (sedimentation, pollution, fishing) are, to a certain extent, large-scale cross-boundary problems that may best be alleviated by management on a lake-basin scale (Pendleton and van Breda, 1994). It is also true, however, that the effects of fishing and sedimentation can be relatively localized. In many cases the fisheries operate on fish stocks of limited distribution and motility, and sedimentation occurs in parts of the catchment degraded by deforestation, while other parts of the catchment (particularly in existing terrestrial parks) have few sediment/pollution problems (Coulter and Mubamba, 1994). For a number of practical reasons, focusing conservation efforts on extension of existing terrestrial parks into the littoral and sub-littoral zones of the lake is likely to be a useful conservation strategy. Lateral extension to protect valuable littoral biotopes, such as reed beds, river estuaries and coastal lagoons, may also be desirable. Zoned conservation areas are increasingly being used to reconcile the conflicting demands of exploitation and habitat preservation, and it may be feasible to allow certain exploitative activities, to a limited extent, in areas of parks. In Lake Malawi National Park, enclaves where fishing is permitted exist within the park and serve to allow established fishing communities to continue to exploit pelagic fish from bases within the park, where the littoral fish community is protected from exploitation (Lewis *et al.*, 1986).

Balon (1993), in pointing out that biodiversity is a dynamic quantity – new forms and species are evolving, as well as being lost – makes a plea that we should "attempt to protect the ongoing natural phenomenon of form creation and change from human-induced acceleration or extirpation", rather than simply making inventories of yet undiscovered or diminishing biodiversity. Thus the priority should be habitat and ecosystem protection, rather than efforts to catalogue existing diversity or to protect individual species.

In the marine environment, reserves have recently been the focus of fishery management strategies. Reserves have been created not for tourism, biodiversity conservation or habitat protection (although these are important secondary benefits), but for their potential in enhancing recruitment to exploited fish stocks in the surrounding areas.

BIBLIOGRAPHY

ABE, N. (1988a) Some notes on crabs at Mbemba. *Ecol. Limnol. Tanganyika*, **5** : 61.

ABE, N. (1988b) Preliminary report on the feeding ecology of Mastacembelids in Lake Tanganyika. *Ecol. Limnol. Tanganyika*, **5** : 37.

ABE, N. (1989) Social organization and parental care of *Afromastacembelus platysoma*. *Ecol. Limnol. Tanganyika*, **6** : 38.

ALLISON, E. H. (1996) Estimating fish production and biomass in the pelagic zone of Lake Malawi: a comparison between acoustic observations and predictions based on biomass size–distribution theory. pp. 224–242. In: COX, I. G. (ed.) *Stock Assessment in Freshwater Fisheries.* Oxford: Fishing News Books/Blackwell Scientific.

ALLISON, E. H., DAVIES, A., NGATUNGA, B. P. and THOMPSON, A. B. (1994) A method for studying fish communities in deep lakes using drifting gillnets. *Fish. Res.*, **20** : 87–91.

ALLISON, E. H., IRVINE, K. and THOMPSON, A. B. (in press) Lake flies and the deep-water demersal fish community of Lake Malawi. *J. Fish Biol.*

ALLISON, E. H., PATTERSON, G., IRVINE, K., THOMPSON, A. B. and MENZ, A. (1995a) The pelagic ecosystem. pp. 351–377. In: MENZ, A. (ed.) *The Fisheries Potential and Productivity of the Pelagic Zone of Lake Malawi/Niassa.* Chatham: Natural Resources Institute.

ALLISON, E. H., THOMPSON, A. B., NGATUNGA, B. P. and IRVINE, K. (1995b) The diet and food consumption rates of the offshore fish. pp. 233–278. In: MENZ, A. (ed.) *The Fisheries Potential and Productivity of the Pelagic Zone of Lake Malawi/Niassa.* Chatham: Natural Resources Institute.

ANDERSON, A. N. (1995) Measuring more of biodiversity: genus richness as a surrogate for species richness in Australian ant faunas. *Biol. Conserv.*, **73** : 39–43.

ANGEL, M. V. (1990) Life in the benthic boundary layer: connections to the mid-water and sea floor. *Phil. Trans. R. Soc. Lond. A*, **331** : 15–28.

ANON. (1965) *Fisheries Research Bulletin 1962–63*. Republic of Zambia, Ministry of Lands and Natural Resources, Game and Fisheries Department. Lusaka, Zambia: Government Printer.

BADENHUIZEN, G. R. (1965) Lufuba River – Research Notes. In: MORTIMER, M. A. E. (ed.) *Fisheries Research Bulletin 1963–64*. pp. 11–43. Republic of Zambia, Ministry of Lands and Natural Resources, Game and Fisheries Department. Lusaka, Zambia : Government Printer.

BAILEY, R. M. and STEWART, D. J. (1984) Bagrid catfishes from Lake Tanganyika, with a key and descriptions of new taxa. *Misc. Publ. Mus. Zool. Univ. Michigan*, **168** : 1–41.

BALON, E. K. (1981) Additions and amendments to the classification of reproductive styles in fishes. *Env. Biol. Fish.*, **6** : 377–389.

BALON, E. K. (1993) Dynamics of biodiversity and mechanisms of change: a plea for balanced attention to form creation and extinction. *Biol. Conserv.*, **66** : 5–16.

BANISTER, K. E. and CLARKE, M. A. (1980) A revision of the large Barbus of Lake Malawi with a reconstruction of the history of Southern Africa rift valley lakes. *J. nat. Hist.*, **14** : 438–542.

BARTSCH, J., BRANDER, K., HEATH, M., MUNK, P., RICHARDSON, K. and SVENDSEN, E. (1989) Modelling the advection of herring larvae in the North Sea. *Nature*, **340** : 632–636.

BAUTE, P., PONCIN, P. and MUZIQWA, K. (1992) Preliminary study of the behaviour of *Distichodus sexfasciatus* Boulenger 1897 in aquarium: ethogram, social behaviour and daily rhythm of feeding activity. *Cah. Ethol. Appl.*, **12** : 509–518.

BAYONA, J. D. R. (1988) A review of the biological productivity and fish predation in Lake Tanganyika. pp. 1–17. In: LEWIS, D. (ed.) *Predator–Prey Relationships, Population Dynamics and Fisheries Productivities of Large African Lakes.* Rome: FAO/CIFA Occasional Papers No. 15.

BAYONA, J. D. R. (1991a) Some aspects of the biology of Kuhe, *Boulengerochromis microlepis*, in the Kigoma region, eastern coast of Lake Tanganyika. *Afr. Stud. Mon.*, **12** : 63–74.

BAYONA, J. D. R. (1991b) Species composition and some observations on the littoral fishes based on beach-seining in the Kigoma region, eastern coast of Lake Tanganyika. *Afr. Stud. Mon.*, **12** : 75–86.

BEADLE, L. C. (1981) *The Inland Waters of Tropical Africa*. Harlow, UK: Longman.

BOOTSMA, H. A. and HECKY, R. E. (1993) Conservation of the African Great Lakes: a limnological perspective. *Conserv. Biol.*, **7** : 644–655.

BOULENGER, G. A. (1905) A list of the freshwater fishes of Africa. *Annals and Magazine of Natural History, London*, 7th Series, **16** : 36–60.

BRICHARD, P. (1978) *Fishes of Lake Tanganyika*. Neptune City, USA: Tropical Fish Hobbyist Publications.

BRICHARD, P. (1989) *Pierre Brichard's Book of Cichlids and all Other Fishes of Lake Tanganyika*. Neptune City, USA: Tropical Fish Hobbyist Publications.

CAPORT, A. (1952) *Le Milieu Géographique et Géophysique. Explorations Hydrobiologique du Lac Tanganyika*, **3** : 39–67. Brussels: Institut Royal de Sciences Naturelles de Belgique.

CHAPMAN, D. W., BAYONA, J. and ELLIS, C.(1974) *Preliminary Analysis of Test Fishing and Limnological Sampling in Tanzanian Waters of Lake Tanganyika (Tanzania)*. FAO Field Document FI:DP/URT/71/012/7. Rome: FAO.

CHAPMAN, D. W. and BAZIGOS, G. P. (1974) *Preliminary Appraisal of the Canoe Fishery of Lake Tanganyika (Tanzania)*. FAO Field Document FI:DP/URT/71/012/35. Rome: FAO.

CHAPMAN, D. W. and VAN WELL, P. (1974a) Growth and mortality of *Stolothrissa tanganicae*. *Trans. Am. Fish. Soc.*, **107** : 26–35.

CHAPMAN, D. W. and VAN WELL, P. (1974b) Observations on the biology of *Luciolates stappersi* in Lake Tanganyika (Tanzania) *Trans. Am. Fish. Soc.*, **107** : 567–573.

CHENE, G. (1975) *Etude des Problèmes Relatifs aux Fluctuations Piscicoles du Lac Tanganika. Mem. lic. Univ. Liège, Belgium.*

COCQUYT, C., CALJON, A. and VYVERMAN, W. (1991) Seasonal and spatial aspects of phytoplankton along the north-eastern coast of lake Tanganyika. *Ann. Hydrobiol.*, **27** : 215–225.

COENEN, E. J. (1993) Classification of Lake Tanganyika shoreline. *LTR Newsletter*, **4** : 8–9.

COENEN, E. J. (1994a) Lake Tanganyika fisheries : an update. *LTR Newsletter*, **11** : 8–9.

COENEN, E. J. (1994b) *Field Guide containing Maps of the Lake Tanganyika Shoreline*. FAO–GCP/RAF/271/FIN–FM/01. Rome: FAO/FINNIDA.

COENEN, E. J. and NIKOMEZE, E. (1994) Lake Tanganyika Catch Assessment Surveys, Burundi, 1992–93. *LTR Newsletter*, **10** : 7–9.

COHEN, A. S. (1989) The taphonomy of gastropod shell accumulations in large lakes: an example from Lake Tanganyika, Africa. *Paleobiology*, **15** : 26–45.

COHEN, A. S. (1991a) Patterns and controls of biodiversity within Lake Tanganyika. pp. 50–53. In: COHEN, A. S. (ed.) *First International Conference on the Conservation and Biodiversity of Lake Tanganyika*.

COHEN, A S. (ed.) (1991b) *Report on the First International Conference on the Conservation and Biodiversity of Lake Tanganyika*. WWF Biodiversity Support Programme/Nature Conservancy and World Resources Institute.

COHEN, A. S. (1992) Criteria for developing viable underwater natural reserves in Lake Tanganyika. *Mitt. int. Ver. Limnol.*, **28** : 109–116.

COHEN, A. S. (1994) Extinction in ancient lakes: biodiversity crises and conservation 40 years after J. L. Brooks. *Arch. Hydrobiol. Beih. Ergebn. Limnol.*, **44** : 451–479.

COHEN, A. S., BILLS, R., COCQUYT, C. Z. and CALJON, A. G. (1993) The impact of sediment pollution on biodiversity in Lake Tanganyika. *Conserv. Biol.*, **7** : 667–677.

COHEN, A. S., KAUFMAN, L. and OGUTU–OHWAYO, R. (in press). Anthropogenic threats, impacts and conservation strategies in the African Great Lakes – a review. In: JOHNSON, T. A. and ODADA, E. (eds) *The Limnology, Climatology and Palaeo-climatology of the East African Lakes*. London: Gordon and Breach.

COULTER, G. W. (1961) Lake Tanganyika Research. *Ann. Rep. JFRO, Zambia,* **10** : 7–30.

COULTER, G. W. (1966) *Hydrobiological Processes and the Deepwater Fish Community in Lake Tanganyika.* PhD thesis, Queens University, Belfast.

COULTER, G. W. (1967) Low apparent oxygen requirement of deep water fishes in Lake Tanganyika. *Nature*, **215** : 317–318.

COULTER, G. W. (1968) Thermal stratification in the deep hypolimnion of Lake Tanganyika. *Limnol. Oceanogr.*, **13** : 385–387.

COULTER, G. W. (1970) Population changes within a group of fish species in Lake Tanganyika following their exploitation. *J. Fish Biol.*, **2** : 329–353.

COULTER, G. W. (1976) The biology of *Lates* species (Nile perch) in Lake Tanganyika, and the status of the pelagic fishery for *Lates* species and *Luciolates stappersi* (Blgr) *J. Fish Biol.*, **9** : 235–259.

COULTER, G. W. (1991a) *Lake Tanganyika and its Life*. London: British Museum (Natural History)/Oxford University Press.

COULTER, G. W. (1991b) Composition of the flora and fauna. pp. 200–274. In: COULTER, G. W. (ed.) *Lake Tanganyika and its Life*. London: British Museum (Natural History)/Oxford University Press.

COULTER, G. W. (1991c) Zoogeography, affinities and evolution, with special regard to the fish. pp. 275–307. In: COULTER, G. W. (ed.) *Lake Tanganyika and its Life*. London: British Museum (Natural History)/Oxford University Press.

COULTER, G. W. (1991d) Fisheries. pp. 139–150. In: COULTER, G. W. (ed.) *Lake Tanganyika and its Life*. London: British Museum (Natural History)/Oxford University Press.

COULTER, G. W. (1991e) Pelagic fish. pp. 111–138. In: COULTER, G. W. (ed.) *Lake Tanganyika and its Life*. London: British Museum (Natural History)/Oxford University Press.

COULTER, G W. (1991f) The benthic fish community. pp. 151–199. In: COULTER, G. W. (ed.) *Lake Tanganyika and its Life*. London: British Museum (Natural History)/Oxford University Press.

COULTER, G. W. and MUBAMBA, R. (1993) Conservation in Lake Tanganyika, with special reference to underwater Parks. *Conserv. Biol.*, **7** : 678–685.

COULTER, G. W. and MUBAMBA, R. (1994) Response to Pendleton and van Breda [Underwater parks may not be the best conservation tool for Lake Tanganyika]. *Conserv. Biol.*, **8** : 331–333.

DAGET, J., GOSSE, J. P. and THYS VAN DEN AUDENAERDE, D. F. E. (eds) (1984) *Checklist of the Freshwater Fishes of Africa*, Vol. 1. Paris/Tervuren: ORSTOM/MRAC.

DAGET, J., GOSSE, J. P. and THYS VAN DEN AUDENAERDE, D. F. E. (eds) (1986a) *Checklist of the Freshwater Fishes of Africa*, Vol. 2. Paris/Tervuren: ORSTOM/MRAC.

DAGET, J., GOSSE, J. P. and THYS VAN DEN AUDENAERDE, D. F. E. (eds) (1986b) *Checklist of the Freshwater Fishes of Africa*, Vol. 3. Paris/Tervuren: ORSTOM/MRAC.

DAGET, J., GOSSE, J. P. and THYS VAN DEN AUDENAERDE, D. F. E. (eds) (1991) *Checklist of the Freshwater Fishes of Africa*, Vol. 4. Paris/Tervuren: ORSTOM/MRAC.

DAGET, J, J P GOSSE, G C TEUGELS and D F E THYS VAN DEN AUDENAERDE (1991) *Checklist of the Freshwater Fishes of Africa* (Cloffa).

DE VOS, L. and SNOEKS, J. (1994) The non-cichlid fishes of the Lake Tanganyika basin. pp. 391-405. In: MARTENS, K., GODDEERIS, B. and COULTER, G. (eds) *Speciation in Ancient Lakes. Arch. Hydrobiol. Beih. Ergebn. Limnol.*, No. 44.

DEELSTRA, H. (1977) Organochlorine insecticide levels in various fish species in Lake Tanganyika. *Med. Fac. Landbouw. Rijksuniv. Gent, Belgium*, **42** : 869–882.

DUMONT, H. J. (1994a) Ancient lakes have simplified pelagic food webs. *Ergebn. Limnol. Archiv. fur Hydrobiol.*, **44** : 223–234.

DUMONT, H. J. (1994b) The distribution and ecology of the fresh and brackish water medusae of the world. *Hydrobiol.*, **272** : 1–12.

DUMONT, H. J. (1994c) On the diversity of the Cladocera in the tropics. *Hydrobiol.*, **272** : 27–38.

ECCLES, D. H. (1986) Is speciation of demersal fishes in lake Tanganyika restrained by physical limnological conditions? *Biol. J. Linn. Soc.*, **29** : 115–122.

ECCLES, D. H. (1992) *Field Guide to the Freshwater Fishes of Tanzania.* FAO Species Identification Sheets for Fishery Purposes, Project URT/87/016. Rome: FAO.

ELLIS, C. M. A. (1971) The size at maturity and breeding seasons of sardines in southern Lake Tanganyika. *Afr. J. Trop. Hydrobiol. Fish.*, **1** : 59–66.

ELLIS, C. M. A. (1978) Biology of *Luciolates stappersi* in Lake Tanganyika (Burundi). *Trans. Am. Fish. Soc.*, **107** : 557–566.

ENDERLEIN, H. O. (1976) *Biological Sampling Survey of the Traditional and Artisanal Fisheries, Lake Tanganyika, Burundi.* UN FAO REP. F1: DP/BD1/70 508/8 : 1–7. Rome: FAO.

ENOKI, A., MAMBONA, W. B. and MUKIRANIA, M. (1987) General survey of fisheries in north-western part of Lake Tanganyika. *Ecol. Limnol. Tanganyika*, **5** : 98–101.

FAO (1978a) *Fishery Biology and Stock Assessment.* FAO Technical Report FI:DP/URT/71/012/1. Rome: FAO.

FAO (1978b) *Expert Consultation on Management of Multispecies Fisheries. Some Scientific Problems of Multispecies Fisheries.* FAO Fisheries Technical Paper No. 181. Rome: FAO.

FRYER, G. and ILES, T. D. (1972) *The Cichlid Fishes of the Great Lakes of Africa.* Edinburgh: Oliver and Boyd.

GASTON, K. J. and WILLIAMS, P. H. (1993) Mapping the world's species – the higher taxon approach. *Biodiversity Lett.*, **1** : 2–8.

GOLDSCHMIDT, T., WITTE, F. and DE VISSER, J. (1993) Cascading effects of the introduced Nile perch on the detritivorous/phytoplanktivorous species in the sublittoral areas of Lake Victoria. *Conserv. Biol.*, **7** : 686–700.

GREBOVAL, D., BELLEMANS, M. and FRYD, M. (1994) *Fisheries Characteristics of the Shared Lakes of the East African Rift.* CIFA Technical Paper No. 24. Rome: FAO.

GREENWOOD, P. H. (1961) A revision of the genus *Dinotopterus* Blgr (Pisces : Clariidae) with notes on the comparative anatomy of the suprabranchial organs in the Clariidae. *Zool. Bull. Br. Mus. Nat. Hist.*, **7** : 217–241.

HAMLEY, J. M. (1975) Review of gill-net selectivity. *J. Fish. Res. Bd Can.*, **32** : 1943–1969.

HANEK, G. and KOTILAINEN, P. (1993) Key results of aerial frame survey. *LTR Newsletter*, **4** : 6–7.

HAY, M. E. (1994) Species as noise in community ecology: do seaweeds block our view of the kelp forest? *Trends Ecol. Evol.*, **9** : 414–416.

HECKY, R. E. (1984) African lakes and their trophic efficiencies: a temporal perspective. pp. 405–448. In: MEYERS, D. G. and STRICKLER, J. R. (eds) *Trophic Interactions Within Aquatic Ecosystems*. AAAS Symposium 85. Washington, DC: Westview Press.

HECKY, R. E. (1991) The pelagic ecosystem. pp. 90–110. In: COULTER, G. W. (ed.) *Lake Tanganyika and its Life*. London: British Museum (Natural History)/Oxford University Press.

HECKY, R. E., BUGENYI, F. W. B., OCHUMBA, P., TALLING, J. F., MUGIDDE, R., GOPHEN, M. and KAUFMAN, L. (1994) Deoxygenation of the deep water of Lake Victoria, East Africa. *Limnol. Oceanogr.*, **39** : 1476–1481.

HECKY, R. E., SPIGEL, R. H. and COULTER, G. W. (1991) The nutrient regime. pp. 76–89. In: COULTER, G. W. (ed.) *Lake Tanganyika and its Life*. London: British Museum (Natural History)/Oxford University Press.

HILL, A. E. (1990) Pelagic dispersal of Norway lobster *Nephrops norvegicus* larvae examined using an advection–diffusion–mortality model. *Mar. Ecol. Prog. Ser.*, **64** : 217–229.

HJORT, J. (1914) Fluctuations in the great fisheries of northern Europe viewed in the light of biological research. *Rapp. p–V. Reun. Cons. int. Explor. Mer.*, **20** : 1–228.

HORI, M. (1983) Feeding ecology of thirteen species of *Lamprologus* (Teleostei; Cichlidae) coexisting at a rocky shore of Lake Tanganyika. *Physiol. Ecol. Japan*, **20** : 129–149.

HORI, M. (1987) Mutualism and commensalism in the fish community of Lake Tanganyika. pp. 219–239. In: KAWANO, S., CONNELL, J. H. and HIDAKA, T. (eds) *Evolution and Co–adaptation in Biotic Communities*. Tokyo: University of Tokyo Press.

HORI, M. (1995) Fish community of deep bottom in southern part of Lake Tanganyika. *Ecol. Limnol. Tanganyika*, **9** : 21–22.

HORI, M., GASHAGAZA, M. M., NSHOMBO, M. and KAWANABE, H. (1993) Littoral fish communities in Lake Tanganyika: irreplaceable diversity supported by intricate interactions among species. *Conserv. Biol.*, **7** : 657–666.

HORI, M., ROSSITER, A. and SATO, T. (1989) Abundance and microdistribution of rock-dwelling cichlids at Kasenga, southern part of Lake Tanganyika. *Ecol. Limn. Tanganyika*, **6** : 63–66.

HUTCHINSON, G. E. (1959) Homage to Santa Rosalia or why there are so many kinds of animals. *Am. Nat.*, **93** : 145–159.

JACKSON, P. B. N. (1959) A review of the clariid cat-fishes of Nyasaland with a description of a new genus and seven new species. *Proc. zool. Soc. Lond.*, **132** : 109–128.

JFRO (1962) *Joint Fisheries Research Organisation, Annual Report* No. 10. Lusaka, Northern Rhodesia: Government Printer.

KARINO, K. (1991) Abundance and vertical distribution of fishes on the rocky shore of Kasenga, southern Lake Tanganyika. *Ecol. Limnol. Tanganyika*, **7** : 50–52.

KARINO, K. and KUWAMURA, T. (1989) Temporary disappearance of fishes from the shallow water of Kasenga, caused by overcrowding of jellyfish. *Ecol. Limnol. Tanganyika*, **7** : 30.

KAWABATA, M. and MIHIGO, N. Y. K. (1982) Littoral fish fauna near Uvira, northwestern end of Lake Tanganyika. *Afr. Stud. Monogr.*, **2** : 133–143.

KAWANABE, H. (ed.) (1981). *Ecological and Limnological Study on Lake Tanganyika and its Adjacent Regions*, Vol. 1. Kyoto: Kyoto University Press.

KAWANABE, H. (ed.) (1983) *Ecological and Limnological Study on Lake Tanganyika and its Adjacent Regions*, Vol. 2. Kyoto: Kyoto University Press.

KAWANABE, H. (ed.) (1985) *Ecological and Limnological Study on Lake Tanganyika and its Adjacent Regions*, Vol. 3. Kyoto: Kyoto University Press.

KAWANABE, H. (ed.) (1988) *Ecological and Limnological Study on Lake Tanganyika and its Adjacent Regions*, Vol. 5. Kyoto: Kyoto University Press.

KAWANABE, H. (ed.) (1989) *Ecological and Limnological Study on Lake Tanganyika and its Adjacent Regions*, Vol. 6. Kyoto: Kyoto University Press.

KAWANABE, H. (ed.) (1991) *Ecological and Limnological Study on Lake Tanganyika and its Adjacent Regions*, Vol. 7. Kyoto: Kyoto University Press.

KAWANABE, H. (ed.) (1993) *Ecological and Limnological Study on Lake Tanganyika and its Adjacent Regions*, Vol. 8. Kyoto: Kyoto University Press.

KAWANABE, H. (ed.) (1995) *Ecological and Limnological Study on Lake Tanganyika and its Adjacent Regions*, Vol. 9. Kyoto: Kyoto University Press.

KAWANABE, H. and NAGOSHI, M. (eds) (1987) *Ecological and Limnological Study on Lake Tanganyika and its Adjacent Regions*, Vol. 4. Kyoto: Kyoto University Press.

KENDALL, R. L. (1973a) *Basic Data in Support of a Report: The Benthic Gill Net Fishery of Lake Tanganyika, Zambia, 1960–1972.* Zambia: Department of Wildlife and Fisheries (mimeo).

KENDALL, R. L. (1973b) *The Benthic Gill Net Fishery of Lake Tanganyika, Zambia, 1960–1972.* Zambia: Department of Wildlife and Fisheries (mimeo).

KIMBADI, S. (1989) Preliminary report on relation among body size, clutch size and egg size of shrimps in the Northwestern part of Lake Tanganyika. *Ecol. Limnol. Tanganyika*, **6** : 45–47.

KIMBADI, S. (1991) Comparison in reproductive biology of some Limnocaridinae shrimps in the Northwestern part of Lake Tanganyika. *Ecol. Limnol. Tanganyika*, **7** : 93–95.

KIMBADI, S. (1995) Preliminary report on shrimps in the south of Lake Tanganyika. *Ecol. Limnol. Tanganyika*, **9** : 35–36.

KIMURA, S. (1995) Growth of the clupeid fishes, *Stolothrissa tanganicae* and *Limnothissa miodon*, in the Zambian waters of Lake Tanganyika. *J. Fish Biol.*, **47** : 569–575.

KINOSHITA, I. and TSHIBANGU, K. K. (1989) Identification of larvae and juveniles of four centropomids from Lake Tanganyika. *Ecol. Limnol. Tanganyika*, **6** : 14.

KOHDA, M. (1987) Territory size of *Tropheus moorii* and *Petrochromis trewavasea* in relation to habitat depth. *Ecol. Limnol. Tanganyika*, **4** : 22–23

KONDO, T. (1985) Breeding biology of a small unionid mussel, *Moncetia lavigeriana*. *Ecol. Limnol. Tanganyika*, **3** : 28–29.

KONDO, T. (1987) Reproductive biology of a small bivalve *Grandidieria burtoni*. *Ecol. Limnol. Tanganyika*, **4** : 122.

KONDO, T. and ABE, N. (1989) Habitat preference, food habits and growth of juveniles of three *Lates* species. *Ecol. Limnol. Tanganyika*, **6** : 15–16.

KONINGS, A. (1988) *Tanganyika Cichlids. Verduin Cichlids*. Cichlid Press.

KONINGS, A. and DIECKHOFF, H. W. (1992) *Tanganyika Secrets*. Cichlid Press.

KUUSIPALO, L. (1994) *Evaluation des Structures Génétiques des Populations de Poisson Pélagique au Lac Tanganyika*. FAO GCP/RAF/271/FIN–TD/23. Rome: FAO.

KUWAMURA, T. (1987) Distribution of fishes in relation to the depth and substrate at Myako, east-middle coast of Lake Tanganyika. *Afr. Stud. Mongr.*, **7** : 1–14.

KUWAMURA, T. (1988) Biparental mouth brooding and guarding in a Tanganyikan cichlid *Haplotaxon microlepis*. *Jpn. J. Ichthyol.*, **35** : 62–68.

KUWAMURA, T. and KARINO, K. (1991) Littoral fish fauna along the Southern coast of Lake Tanganyika, Zambia, with special reference to the family Cichlidae. *Ecol. Limnol. Tanganyika*, **7** : 14–17.

KWETUENDA, M. K. (1983) L'importance des lagunes cotières dans le phenomène de reproduction de la Faune ichthyologique à l'estuaire de la Ruzizi, Tanganika. *Ecol. Limnol. Tanganyika*, **2** : 47–49.

KWETUENDA, M. K. (1987) Fish fauna in temporal lagoons at northern end of Lake Tanganyika. *Ecol. Limnol. Tanganyika*, **4** : 67–72.

LELOUP, E. (1953) Gastéropodes. Résultats scientifiques de l'exploration hydrobiologique du Lac Tanganyika (1946–1947). *Institut Royal des Sciences Naturelles de Belgique*, **34** (4) : 1–272.

LEVEQUE, C. (1995) Role and consequences of fish diversity in the functioning of African freshwater ecosystems: a review. *Aquat. Living Resour.*, **8** : 59–78.

LEWIS, D., REINTHAL, P. and TRENDALL, J. (1986) *A Guide to the Fishes of Lake Malawi National Park.* Gland/WWF.

LINDQUIST, O. V. and MIKKOLA, H. (1989) *Lake Tanganyika: Review of Limnology, Stock Assessment, Biology of Fishes and Fisheries.* Report GCP/RAF/229/FIN for Regional Project for the Management of Fisheries, Lake Tanganyika. Rome: FAO.

LOMOLINO, M. V. (1994) An evaluation of alternative strategies for building networks of nature reserves. *Biol. Conserv.*, **69** : 243–249.

LOWE-McCONNELL, R. H. (1987) *Ecological Studies in Tropical Fish Communities.* Cambridge: Cambridge University Press.

LOWE-McCONNELL, R. H. (1993) Fish faunas of the African Great Lakes: origins, diversity and vulnerability. *Conserv. Biol.*, **7** : 634–643.

LOWE-McCONNELL, R. H. (1994) Ecological and behaviour studies of cichlids. *Arch. Hydrobiol. Beih. Ergebn. Limnol.*, **44** : 335–345.

LOWE-McCONNELL, R. H., CRUL, R. C. M. and ROEST, F. C. (eds) (1992) Symposium on resource use and conservation of the African Great Lakes, Bujumbura, 1989. *Mitt. int. Ver. Limnol.*, **23** : 1– 128.

MANN, M. J. and NGOMIRAKIZA, M. (1973) Evaluations of the pelagic resources in the Burundi Waters of Lake Tanganyika and the evolution of the fisheries. *Afr. J. Trop. Hydrobiol. Fish. Spec.,* (2) : 145–142.

MARLIER, G. (1957) Le Ndagala, poisson pélagique du Lac Tanganika. *Bull. agric. Congo belge*, **48** : 409–422.

MARLIER, G. (1962) Genera des Trichoptères en Afrique. Annales du Museé Royale de l'Afrique centrâle, *Sciences Zoologiques*, **109** : 1–261.

MARSHALL, B. (1993) Biology of the African clupeid *Limnothrissa miodon* with reference to its small size in artificial lakes. *Rev. Fish Biol. Fisheries*, **3** : 17–38.

MARTENS, K. (1985) *Tanganyikacypridopsis* gen. n. (Crustacea, Ostracoda) from Lake Tanganyika. *Zool. Scr.* **14** : 221–230.

MARTENS, K. (1994) Ostracodes in ancient lakes. *Arch. Hydrobiol. Beih. Ergebn. Limnol.*, **44** : 203–222.

MARTENS, K., COULTER, G. W. and GODDEERIS, B. (eds) (1994) Speciation in ancient lakes. *Arch. Hyrobiol. Beih. Ergebn. Limnol.*, **44** : 1–508.

MASHIKO, K., KAWABATA, S. and OKINO, T. (1991) Reproductive and populational characteristics of a few caridean shrimps collected from Lake Tanganyika, East Africa. *Arch. Hyrobiol. Beih. Ergebn. Limnol.*, **22** : 69–78.

MATSUDA, H., HORI, M. and ABRAMS, P. A. (1994) Effects of predator-specific defence on community complexity. *Evol. Ecol.*, **8** : 628–638.

MATTHES, H. (1960) *Note sur la Reproduction des Poissons au Lac Tanganika.* Publications Conseil Scientifique pour l'Afrique au sud du Sahara No. 63. London: Kikuyu.

MATTHES, H. (1967) Preliminary investigations into the biology of the Lake Tanganyika Clupeidae. *Fish. Res. Bull. Zambia*, **4** : 39–45.

MATTHES, H. (1975) A key to the families and genera of freshwater fishes of Tanzania. *Afr. J. Trop. Hydrobiol. Fish.*, **4** : 166–183.

MENZ, A. (ed.) (1995) *The Fisheries Potential and Productivity of the Pelagic Zone of Lake Malawi/Niassa*. Chatham: Natural Resources Institute.

MICHEL, A. E., COHEN, A. S., WEST, K., JOHNSTON, M. R. and KAT, P. W. (1992) Large African lakes as natural laboratories for evolution: examples from the endemic gastropod fauna of Lake Tanganyika. *Mitt. int. Ver. Limnol.*, **23** : 85–99.

MIHIGO, N. K. (1983) Distribution of adult fishes and description of fish larvae in the north-western part of Lake Tanganyika. *Ecol. Limnol. Tanganyika*, **2** : 20–23.

MOLSA, H. (ed.) (1995) *Abstracts of a Symposium on Lake Tanganyika Research*, September 11–15, 1995, Kuopio, Finland. Kuopio University Publications.

MOREAU, J., MUNYANDORERO, J. and NYAKAGENI, B. (1991) Evaluation des parametres demographiques chez *Stolothrissa tanganyikae* et *Limnothrissa miodon* du Lac Tanganyika. *Verh. int. Verein. theoret. angew. Limnol., Stuttgart,* **24** : 2552–2558.

MOREAU, J. and NYAKAGENI, B. (1988) Les relations trophiques dans la zone pelagique du Lac Tanganyika (secteur Burundi): essai d'evaluation. *Rev. Hydrobiol. Trop.*, **23** : 157–164.

MOREAU, J. and NYAKAGENI, B. (1992) *Luciolates stappersi* in Lake Tanganyika. Demographical status and possible recent variations assessed by length frequency distributions. *Hydrobiol.*, **232** : 57–64.

MOREAU, J., NYAKAGENI, B., PEARCE, M. and PETIT, M. (1993) Trophic relationships in the pelagic zone of Lake Tanganyika (Burundi Sector). *ICLARM Conf. Proc.*, **26** : 138–143.

MULIMBWA, N. (1991) Life cycles, growth and spawning season of ndagala, *Stolothrissa tanganicae* and *Limnothrissa miodon*, in the northwestern part of Lake Tanganyika. *Ecol. Limnol. Tanganyika*, **7** : 45–46.

MULIMBWA, N. and SHIRAKIHARA, K. (1994) Growth, recruitment and reproduction of sardines (*Stolothrissa tanganicae* and *Limnothrissa miodon*) in northwestern Lake Tanganyika. *Tropics*, **4** : 57–67.

NAEEM, S., THOMPSON, L. J., LAWLER, S. P., LAWTON, J. H. and WOODFIN, R. M. (1994) Declining biodiversity can alter the performance of ecosystems. *Nature*, **368** : 734–737.

NAKAYA, K. (1991) Fishes in the north-western waters of Lake Tanganyika, collected from July through September, 1990. *Ecol. Limnol. Tanganyika*, **7** : 81–83.

NARITA, T. (1987) Distribution and reproductive characters of atyid shrimps. *Ecol. Limnol. Tanganyika*, **4** : 114.

NARITA, T. (1995a) Atyid shrimp fauna in Zambian waters around Mpulungulu. *Ecol. Limnol. Tanganyika*, **9** : 44–45.

NARITA, T (1995b) Protecting effect of nursery territory of *Neolamprologus moorei* for atyid shrimp. *Ecol. Limnol. Tanganyika*, **9** : 46–47.

NDARO, S. G. M. (1990) *Study of the Inshore Water Cichlid Fish Potential of Lake Tanganyika around Kigoma, Tanzania*. Occasional Paper No. 3. Wageningen: International Agriculture Centre, Fisheries and Aquaculture Unit.

NDARO, S. G. M. (1992) Nearshore fish resources and fisheries around Kigoma, eastern coast of Lake Tanganyika. *ICLARM Naga*, **15** : 35–38.

NISHIDA, M. (1988) Genetic variability of the clupeid fishes, *Stolothrissa tanganicae* and *Limnothrissa miodon*, in Lake Tanganyika. *Ecol. Limnol. Tanganyika*, **5** : 38.

OCHI, H., YANAGISAWA, Y. and OMORI, K. (1995) Intraspecific brood-mixing of the cichlid fish *Perissodus microlepis* in Lake Tanganyika. *Env. Biol. Fish.*, **43** : 201–206.

PAULY, D. (1980) On the interrelationships between natural mortality, growth parameters and mean environmental temperature in 175 fish stocks. *J. Cons. int. Explor. Mer.*, **39** : 175–192.

PAYNE, A. I. (1986) *The Ecology of Tropical Lakes and Rivers*. Chichester: Wiley.

PEARCE, M. J. (1985a) *A Description and Stock Assessment of the Pelagic Fishery in the South-East arm of the Zambian Waters of Lake Tanganyika.* Zambia: Department of Fisheries.

PEARCE, M. J. (1985b) *The Deepwater Demersal Fish in the South of Lake Tanganyika.* Zambia: Department of Fisheries.

PEARCE, M. J. (1985c) *Some Effects of* Lates *Species on Pelagic and Demersal Fish in Zambian Waters of Lake Tanganyika.* FAO/CIFA Symposium SAWG/85/W/P2. Rome: FAO.

PEARCE, M. J. (1988) Some effects of *Lates* spp. on pelagic and demersal fish in Zambian waters of Lake Tanganyika. In : LEWIS, D. (ed.) *Predator–Prey Relationships, Population Dynamics and Fisheries Productivities of Large African Lakes.* FAO/CIFA Occasional Paper No. 15. Rome: FAO.

PEARCE, M. J. (1989) Preliminary report on stomach content analysis of a commercially important pelagic fish, *Lates stappersi. Ecol. Limnol. Tanganyika,* **6** : 80–81.

PEARCE, M. J. (1990) *Thirty Years of Exploitation of the Pelagic Fish Stocks in the Zambian Waters of Lake Tanganyika.* Occasional Paper No. 3. Wageningen: International Agriculture Centre, Fisheries and Aquaculture Unit.

PEARCE, M. J. (1991a) Food and gonad maturity of *Luciolates stappersi* in southern Lake Tanganyika. *Ecol. Limnol. Tanganyika,* **7** : 31–34.

PEARCE, M. J. (1991b) A note on the importance of prawns to the ecology of southern lake Tanganyika. *Ecol. Limnol. Tanganyika,* **7** : 72.

PEARCE, M. J. (1991c) Zooplankton abundance in the pelagic waters at the south of lake Tanganyika. *Ecol. Limnol. Tanganyika,* **7** : 73.

PEARCE, M. J. (1995) Effects of exploitation on the pelagic fish community in the south of Lake Tanganyika. pp. 425–442. In: PITCHER, T. J. and HART, P. J. B. (eds) *The Impacts of Species Changes in African Lakes.* London: Chapman & Hall.

PENDLETON, L. H. and VAN BREDA, A. (1994) Underwater parks may not be the best conservation tool for Lake Tanganyika. *Conserv. Biol.,* **8** : 330–331.

PETIT, P. and KIYUKU, A. (1995) Changes in the pelagic fisheries of northern Lake Tanganyika during the 1980s. pp. 443–455. In: PITCHER, T. J. and HART, P. J. B. (eds) *The Impacts of Species Changes in African Lakes.* London: Chapman & Hall.

PETIT, P. and NYAKAGENI, B. (1994) *Les Espèces Piscicoles du Secteur Burundais du Lac Tanganika.* Ecologie des poissons tropicaux: Projet ECOTONES-UNESCO, 1st edition, Mai 1994. Burundi: Université du Burundi.

PHIRI, H. (1991a) Preliminary report on the study of the food of Lake Tanganyika Clupeids. *Ecol. Limnol. Tanganyika,* **7** : 66–67.

PHIRI, H. (1991b) Deep water fish community of southern Lake Tanganyika. *Ecol. Limnol. Tanganyika,* **7** : 53–54.

POLL, M. (1952) Poissons des rivières de la region des lacs Tanganika et Kivu recueillis par G. Marlier. *Rev. Zool. Bot. afr.,* **46** : 221–237.

POLL, M. (1953) *Poissons Non-Cichlidae. Resultats Scientifique de l' Exploration Hydrobiologique du Lac Tanganika.* Brussels: Institut Royal des Sciences Naturelles de Belgique. No. 3 (5A).

POLL, M. (1956) *Exploration Hydrobiologique du Lac Tanganika.* Brussels: Institut Royal des Sciences Naturelles de Belgique.

POLL, M. (1957) Les genres des poissons d'eau douce de l'Afrique. *Annls Mus. r. Congo belge,* **54** : 1–191.

POLL, M. (1986) Classification des Cichlidae du lac Tanganika: Tribus, Genres et espèces. *Mém. Cl. Sci. Acad. Belgique,* **45**(8) : 163.

ROBERTS, T. R. (1975) Geographical distribution of African freshwater fishes. *Zool J. Linn. Soc.*, **57** : 249–319.

ROEST, F. C. (1988) Predator–prey relations in northern Lake Tanganyika and fluctuations in the pelagic fish stocks. pp. 104–129. In: LEWIS, D. (ed.) *Predator–Prey Relationships, Population Dynamics and Fisheries Productivities of Large African Lakes.* FAO/CIFA Occasional Paper No. 15. Rome: FAO.

ROEST, F. C. (1992) The pelagic fisheries resources of Lake Tanganyika. *Mitt. int. Verein. theoret. angew. Limnol.*, **23** : 11–15.

ROSSITER, A. (1993) Species, species, everywhere. *Env. Biol. Fish.*, **37** : 97–101.

SAETERSDAL, M., LINE, J. M. and BIRKS, H. J. B. (1993) How to maximize biological diversity in nature reserve selection: vascular plants and breeding birds in deciduous woodlands, Western Norway. *Biol. Conserv.*, **66** : 131–138.

SAHIRI, C. (1991) Chemical pollution of Lake Tanganyika in the vicinity of Bujumbura, Burundi. pp. 108–109. In: COHEN, A. S.(ed.) *Report on the First International Conference on the Conservation and Biodiversity of Lake Tanganyika.* USA Biodiversity Support Programme.

SALONEN, K. and SARVALA, J. (1994) *Sources of Energy for Secondary Production in Lake Tanganyika. Objectives, Approaches and Initial Experiences.* FAO-GCP/RAF/271/FIN-TD/26. Rome: FAO/FINNIDA.

SATO, T. (1986) A brood parasitic catfish of mouth brooding cichlid fishes in Lake Tanganyika. *Nature*, **323** : 58–59.

SATO, T., YUMA, M., NIIMURA, Y., NAKAI, K., ABE, N., NISHIDA, M., NSHOMBO, M. and YAMAGISHI, S. (1988) Fish fauna around Ubwari Peninsula. *Ecol. Limnol. Tanganyika*, **5** : 14–15.

SEEGERS, L. (1995) Revision of the Kneriidae of Tanzania with description of three new Kneria species (Teleostei: Gonorhynchiformes). *Ichthyol. Explor. Freshwat.*, **6** : 97–128.

SNOEKS, J., RÜBER, L. and VERHEYEN, E. (1994) The Tanganyika problem: comments on the taxonomy and distribution patterns of its cichlid fauna. *Arch. Hydrobiol. Beih. Ergebn. Limnol.*, **44** : 355–372.

SPRULES, W. G. and MUNAWAR, M. (1986) Plankton size-spectra in relation to ecosystem productivity, size and perturbation. *Can. J. Fish. aquat. Sci.*, **43** : 1789–1794.

STURMBAUER, C. and MEYER, A. (1992) Genetic divergence, speciation and morphological stasis in a lineage of African cichlid fishes. *Nature*, **358** : 578–581.

TABORSKY, M. (1984) Broodcare helpers in the cichlid fish *Lamprologus brichardi*: their cost and benefits. *Anim. Behav.*, **32** : 1236–1252.

TAKAHASHI, T., GASHAGAZA, M. M. and NAKAYA, K. (1995) Fishes of Lake Tanganyika around Ubwari peninsula, Zaïre. *Ecol. Limnol. Tanganyika*, **9** : 58–59.

TAKAMURA, K. (1983a) The fish fauna on Mahale mountain coast and a discussion on its characteristics. *Ecol. Limnol. Tanganyika*, **2** : 7–8.

TAKAMURA, K. (1983b) Interspecific relationships between two aufwuchs eaters *Petrochromis polydon* and *Tropheus moorei* (Pisces: Cichlidae) of Lake Tanganyika, with a discussion on the evolution and functions of a symbiotic relationship. *Physiol. Ecol. Japan*, **20** : 59–69.

TAKAMURA, K. (1984) Interspecific relationships of aufwuchs-eating fishes in Lake Tanganyika. *Env. Biol. Fish.*, **10** : 225–241.

THOMPSON, A. B. (1995) Eggs and larvae of *Engraulicypris sardella*. pp. 179–199. In: MENZ, A. (ed.) *The Fisheries Potential and Productivity of the Pelagic Zone of Lake Malawi/Niassa*. Chatham: Natural Resources Institute.

THOMPSON, A. B., ALLISON, E. H. and NGATUNGA, B. P. (1995a) Distribution and breeding biology of offshore pelagic Cyprinidae and catfish in Lake Malawi/Niassa. *Env. Biol. Fish.*

THOMPSON, A. B., ALLISON, E. H. and NGATUNGA, B. P. (1995b) Distribution and breeding biology of offshore pelagic Cichlidae in Lake Malawi/Niassa. *Env. Biol. Fish.*

THOMPSON, K. W., HUBBS, C. and LYONS, B. W. (1977) Analysis of potential environmental factors, especially thermal, which would influence the survivorship of exotic nile perch if introduced into artificially heated reservoirs in Texas. *Texas Parks Wildlife Dept Tech. Ser.*, **22** : 1–37.

TREWAVAS, E., GREEN, J. and CORBET, S. A. (1972) Ecological studies on crater lakes in West Cameroon. Fishes of Barombi Mbo. *J. Zool. Lond.*, **166** : 15–30.

TSHIBANGU, K. K. (1988) Preliminary study on the pelagic fish larvae in the northwestern part of Lake Tanganyika. *Ecol. Limnol. Tanganyika*, **5** : 39.

TSHIBANGU, K. K. and KINOSHITA, I. (1989) Vertical and horizontal distribution of clupeids larvae in the northwestern part of Lake Tanganyika. *Ecol. Limnol. Tanganyika*, **6** :19–20.

VAN MEEL, L. (1952) Le milieu végétal: resultats scientifiques de l'exploration hydrobiologique du Lac Tanganyika (1946–47). *Inst. Royal des Sciences Naturelles de Belgique*, **1** : 51–68.

VERHEYEN, E. (1995) *Wetenschappelijk verslag van een expeditie naar het zuidelijke gedeelte van het Tanganyikameer (Zambia and Tanzania) (Scientific report of an expedition to the southern part of Lake Tanganyika)*. Occasional report (in Dutch).

WEST, G. S. (1907) Report on the freshwater algae including phytoplankton of the third Tanganyikan Expedition (1904–1905). *J. Linn. Soc.*, **38** (264) : 81–197.

WEST, K. and COHEN, A. S. (1991) Morphology and behaviour of crabs and gastropods from Lake Tanganyika, Africa: implications for Lacustrine predator–prey co-evolution. *Evolution*, **45** : 589–607.

WEST, K. and COHEN, A. S. (1994) Predator–prey coevolution as a model for the unusual morphologies of the crabs and gastropods of Lake Tanganyika. *Arch. Hydrobiol. Ergebn. Limnol.*, **44** : 267–283.

WEST, K., COHEN, A. S. and BARON, M. (1991) Morphology and behaviour of crabs and gastropods from Lake Tanganyika, Africa: implications for lacustrine predator–prey coevolution. *Evolution*, **45** : 589–607.

WIJNGAARDEN, C. VAN (1995) *Verspreidingspatronen bij Lamprologus sensu lato (Teleostei, Cichlidea) in het Tanganyikameer (oost Afrika)*. Leuven University Belgium (Distribution patterns in *Lamprologus* sensu lato in Lake Tanganyika). PhD thesis (in Dutch).

WITTE, F., GOLDSCHMIDT, T., WANINK, J., VAN OIJEN, M., GOUDSWAARD, K., WITTE-MAAS, E. and BOUTON, N. (1993) The destruction of an endemic species flock: quantitative data on the decline of the haplochromine cichlids of Lake Victoria. *Env. Biol. Fish.*, **34** : 1–28.

WITTING, L. and LOESCHCKE, V. (1995) The optimisation of biodiversity conservation. *Biol. Conserv.*, **71** : 205–207.

WOELTJE, T. (1995) *Ornamental Fish Trade in the Netherlands*. WWF-Netherlands/Traffic Europe.

WORLD BANK/DANIDA (1994) *Rapid Water Resources Assessment, Vol. 2 Basin Reports, V111 Lake Tanganyika Basin*. Tanzania: Ministry of Water Resources, Energy and Minerals.

WORTHINGTON, E. B. and LOWE-McCONNELL, R. (1994) African lakes reviewed: creation and destruction of biodiversity. *Env. Conserv.*, **21** : 199–213.

WOUTERS, K. and MARTENS, K. (1992) Contribution to the knowledge of Tanganyikan cytheraceans, with the description of Mesocyprideis *nom. nov*. (Crustacea: Ostracoda). *Bull. R. belge. Inst. nat. Biol.*, **62** : 159–166.

WOUTERS, K. and MARTENS, K. (1994) Contribution to the knowledge of the *Cyprideis* species flock (Crustacea: Ostracoda) of Lake Tanganyika, with the description of three new species. *Bull. R. belge. Inst. nat. Biol.*, **64** : 111–128.

YANAGISAWA, Y. (1995) Influence of gastropod shell colonies on diversity of the fish community. *Ecol. Limn. Tanganyika*, **9** : 19.

YANAGISAWA, Y. and NISHIDA, M. (1991) The social and mating system of the maternal mouthbrooder *Tropheus moorii* (Cichlidae) in Lake Tanganyika. *Jpn. J. Ichthyol.*, **38** : 271–282.

YANAGISAWA, Y. and OCHI, H. (1991) Food intake by mouth brooding females of *Cyphotilapia frontosa* (Cichlidae) to feed both themselves and their young. *Env. Biol. Fish.*, **30** : 353–358.

YANAGISAWA, Y. and SATO, T. (1990) Active browsing by mouth brooding females of *Tropheus duboisi* and *Tropheus moorii* (Cichlidae) to feed the young and/or themselves. *Env. Biol. Fish.*, **27** : 43–50.

YUMA, M. (1989) Community structure of benthos-feeding fishes in Lake Tanganyika. *Ecol. Limnol. Tanganyika*, **6** : 39.

YUMA, M. and NAKAI, K. (1987) Brood size and distribution among *Lavigeria* complex (Mollusca: Thiaridae) in Lake Tanganyika. *Ecol. Limnol. Tanganyika*, **4** : 65.

2. IMPACT OF SEDIMENT DISCHARGE AND ITS CONSEQUENCES

2.1 INTRODUCTION

This chapter forms the main body of the baseline review on the Impact of Sediment Discharge and its Consequences, prepared as part of the UNDP-funded project: Pollution Control and Other Measures to Protect Biodiversity in Lake Tanganyika (RAF/92/G32). The baseline review included a full literature review on important aspects of the study, which is presented here. Matters more pertinent to how the project should proceed, and the relevant institutional aspects relating to the project, have been omitted. Persons interested in these other aspects should consult the baseline reviews of the project. The editor has made a number of small changes and additions to the baseline review in the light of comments made by delegates from the four riparian countries (Burundi, Tanzania, Republic of Congo and Zambia) at the project inception workshop held in Dar es Salaam on 25–28 March 1996.

Lake Tanganyika is the largest of the African rift lakes and is also one of the most unusual biotic resources on earth. As yet, Lake Tanganyika receives little legally mandated environmental protection. The most serious immediate problems facing the Lake Tanganyika ecosystem result from overpopulation within the lake basin and the impacts associated with that overpopulation. Excessive suspended sediment input into the lake (which can be linked to basin deforestation), overfishing and pollution are the primary manifestations of this problem. Sedimentation and overfishing, particularly in the northern basin, have led to widespread modification of the lake's ecosystem through local species extinctions and seriously reduced complexity of species interactions within the lake (Cohen, 1991).

2.2 LAND USE IN THE CATCHMENT OF LAKE TANGANYIKA

2.2.1 Introduction

Deforestation, intensification of land use and the spread of unsustainable land use practices are frequent causes of land degradation which may lead to sedimentation and (in places) eutrophication of the lake. Solutions to this form of lake 'pollution' are outside the capabilities of this review, centring as they do on the development of appropriate, and sustainable, improved land use systems and alternative livelihood opportunities for the communities of the area.

Notwithstanding the marked seismic instability of the lake area, recent increases in sedimentation are likely to be associated with large-scale land-use change. The most significant changes are likely to be the advance of subsistence agriculture into former forested areas in response to increasing human populations.

2.2.2 Physiographic setting

Recent studies by Cohen *et al.* (1993) have focused on the impact of increased sediment input from the rivers on the biodiversity of Lake Tanganyika. These authors report that "The lake's basin is undergoing deforestation at an alarming rate; rapid erosion as a consequence of this deforestation is resulting in the discharge of large volumes of sediment into normally clear-water littoral and sublittoral environments". Moreover, they assert that "Deforestation and subsequent rapid erosion are probably the most severe environmental problems currently facing Lake Tanganyika". Analysis of Landsat images has revealed that the problem is most acute at the northern end of the lake (Burundi and northern Republic of Congo) where the population pressure is highest and deforestation approaches 100% of original forested lands (Cohen, 1991; Cohen *et al.*, 1993). Sediment load has been estimated from size of sediment plumes in Landsat imagery and the proportion of deforestation and channel incision in Landsat and aerial photographs, thus deriving maps of light/moderate/highly disturbed regions, maps of deforested areas (<20%, 20–60%, >60%), maps of ostrocode species richness, and maps of main basins (Cohen, 1993). Trefois (1994) has also shown the capability of remote sensing analysis to identify similar sediment inputs and variations in catchment vegetation in Niger.

In the central parts of the lake's drainage basin, between 40 and 60% of the formerly forested lands have been cleared. Forest clearing has been followed by conversion of the former forest lands to grazing or for use in subsistence agriculture. Such clearing may lead to rapid headward erosion, stream incision and gully erosion. The problem is not restricted to Lake Tanganyika (Bruijnzeel, 1990; Cohen, 1991). For example, almost 1000 km to the east, in Tanzania, deforestation and land degradation, accelerating in recent decades, have been identified as the cause of progressively more severe and destructive floods in the Lake Babati catchment in 1964, 1979 and 1990 (Sandström, 1995). Fire is an important indicator of (largely human related) land cover change and, although not the primary agent of deforestation, is certainly an indicator of the loss of forested land to agriculture or settlement and an inhibitor to land recovery. Remote sensing studies of seasonal fire patterns, from National Oceanographic and Atmospheric Administration (NOAA) data, may be a useful tool in establishing zones more likely to be at risk of erosion.

Bizimana and Duchafour (1991) have estimated soil erosion rates of between 20 and 100 t/ha/year in the deforested, steeply sloping and intensively cultivated Nthangwa River basin of northern Burundi. Nearly all of this sediment is discharged into Lake Tanganyika. A consequence of accelerated sediment discharge is the propagation of large deltas. For example, the delta of the Ruzizi River appears to have undergone an order of magnitude increase in its rate of outbuilding subsequent to major deforestation in its drainage area.

2.3 SEDIMENTOLOGY IN THE LAKE TANGANYIKA BASIN AND SEDIMENT INPUTS TO THE LAKE

2.3.1 Introduction

As a precursor to the consideration of sediment inputs to Lake Tanganyika, a brief review is presented of current knowledge of its morphology, bathymetry and water column characteristics. These factors exert significant controls on the dispersal of sediment particles introduced to the lake from inflowing rivers, and on their eventual loci of sedimentation.

2.3.2 Setting and morphometry

Lake Tanganyika is the largest of the Rift Valley lakes of East Africa and the second largest body of fresh water in the world. It occupies the narrow trough of the western branch of the rift system between 3° 30′ and 8° 50′ S and is surrounded by Burundi, Tanzania, Zambia and the Republic of Congo. The lake is *ca* 650 km in length and has an average width of 50 km. The mean elevation of its surface is 773 m above sea level (Edmond *et al.*, 1993). Lake Tanganyika is surpassed in maximum depth only by Lake Baikal, another rift lake. Table 2.1 presents several important morphometric parameters of the lake.

Table 2.1 Morphometric parameters of Lake Tanganyika (after Hutchinson, 1975, and other sources)

Parameter	Value	
Maximum depth	1 470	m
Mean depth	572	m
Surface area	32 000	km^2
Volume	18 940	km^3
Shoreline length	1 900	km

2.3.3 Bathymetry

A bathymetric chart of Lake Tanganyika was produced during the course of a Belgian expedition in 1946–47 (Capart, 1949). This chart, as presented by Edmond *et al.* (1993) shows the water body to be sub-divided into three distinct basins with shallower regions (<500 m in depth) at the northern and southern extremities of the lake. The northern basin, referred to by Edmond *et al.* (1993) as the Kigoma Basin, has a maximum depth of 1310 m. It is separated from the central or Kungwe Basin by a broad sill with a maximum crestal depth of 655 m. This basin extends to a maximum depth of 885 m and is separated from the so-called Kipili Basin to the south by a second, broad sill. The latter basin is the deepest of the lake, extending to the overall maximum depth of 1470 m (Table 2.1).

In their study of the tectonic history of Lake Tanganyika, Rosendahl *et al.* (1986) indicate that four sedimentary basins have been identified in the lake. These are termed, from north to south, the Ruzizi, Kigoma, Kalemie and Marungu-Mpulungu Basins. These basins (Figure 2.1) are separated from each other by ridges of crystalline basement rocks which serve in varying degrees as barriers to sediment spill-over (Rosendahl *et al.*, 1986). This may be an important factor in determining the dispersal and ultimate fate of fine sediments introduced into the lake from influent rivers. The northernmost or Ruzizi Basin was not identified in the 1946–47 survey. Broadly the Kungwe and the Kipili Basins of the latter survey correspond respectively with the Kalemie and Marungu-Mpulungu Basins referred to by Rosendahl *et al.* (1986). Thus the lake lies in multiple grabens and, as such, is a tectonic basin of type 9 as defined by Hutchinson (1975), the most important type of tectonic basin.

The most up to date bathymetry of Lake Tanganyika is presented by Tiercelin and Mondeguer (1991) which is based on the data of Capart (1949) and recent information provided from Project GEORIFT (Elf-Aquitane, France). This chart (Figure 2.2) confirms the earlier findings but shows yet another basin nomenclature for the lake.

2.3.4 Drainage pattern

Over most of its length Lake Tanganyika is bounded by mountains. This is particularly so on its western side where the general altitude is of the order of 2000 m with peaks rising to over 3000 m above sea level at the northern end. The underlying solid geology of its catchment consists predominantly of Precambrian crystalline rocks of Archaean and Proterozoic ages (Dodoma Belt). Localized areas of Mesozoic and Quaternary volcanic rocks are present in Republic of Congo and Burundi.

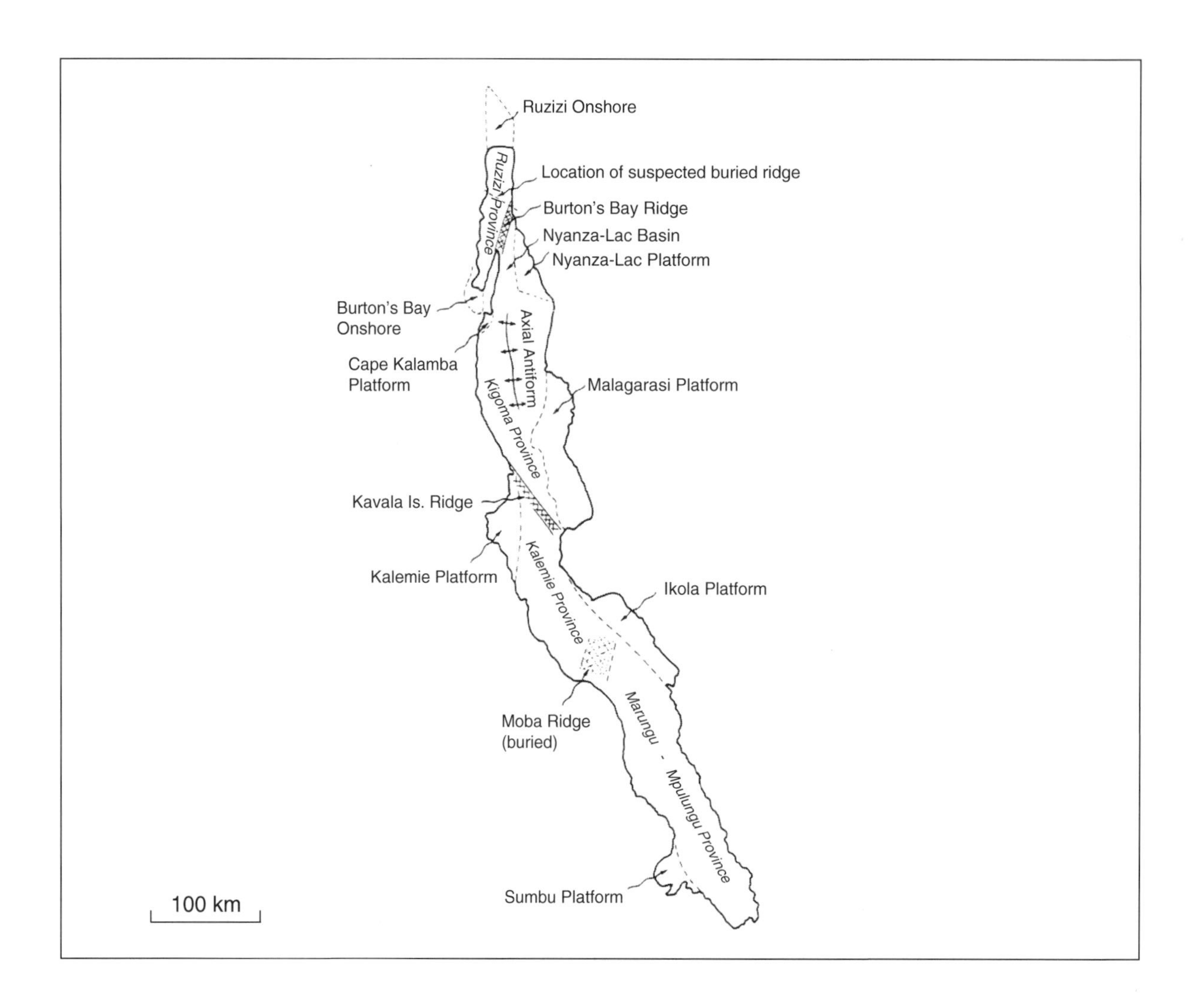

Figure 2.1 Schematic diagram of the lake floor geomorphology of Lake Tanganyika at the present time. Provinces represent modern depositional units. (Source: Rosendahl *et al.*, 1986.)

The drainage pattern of the lake is shown in Figure 2.2. Many rivers enter the lake but it has one outflow, the River Lukuga (referred to as the Lukula by Gasse *et al.*, 1989) which discharges westwards to the River Congo (Republic of Congo) and controls the lake level. Two of the principal influents are the Ruzizi River (also known as the Rusizi River, Figure 2.2), which drains from Lake Kivu and enters Lake Tanganyika at its northern end, and the Malagarasi River which enters the lake on its eastern side (Figure 2.2). Off the mouths of both the Malagarasi River, and the Lugufu River which enters the lake to the south of the former, are located sublacustrine valleys (Hutchinson, 1975). Capart (1949) believed that these were eroded when the lake stood at a much lower level than at present. Recent studies by Gasse *et al.* (1989) have shown that these submerged valleys extend down to 550 m below present lake levels. These valleys may play an important role in constraining the inflow of sediment-laden waters into the lake particularly if the influents are of greater density than the waters of the receptor basin (see below). Not all modern rivers entering the lake lead to submerged channels, however. The Ruzizi River, for example, has two distributaries, referred to by Tiercelin and Mondeguer (1991) as the "Petite and Grande Rusizi", but has only one submerged channel extending from the mouth of the smaller distributary which is known to be older than the larger (Hutchinson, 1975).

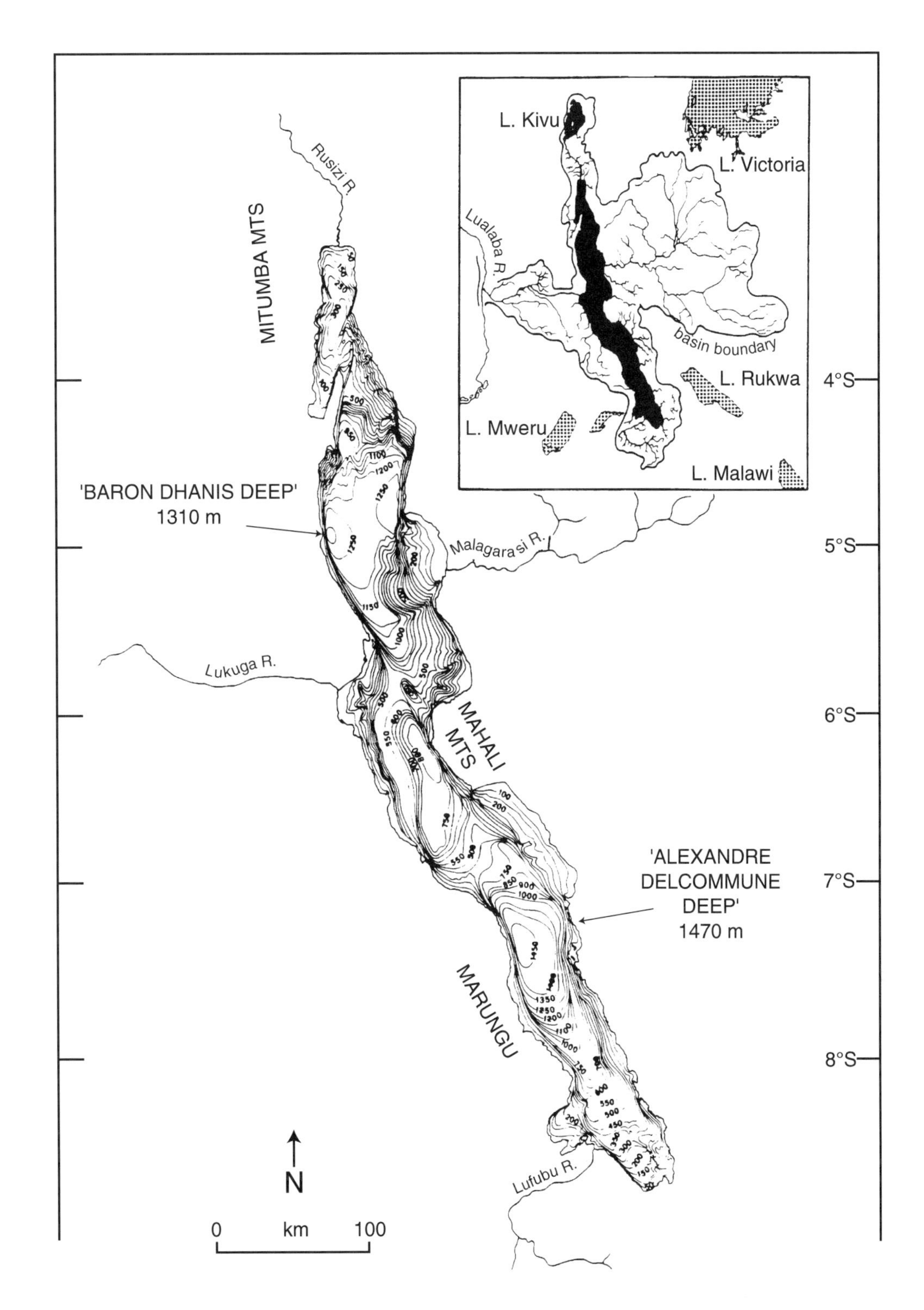

Figure 2.2 Bathymetry and drainage basin of Lake Tanganyika. Bathymetric data are from Capart (1949) and from echo-sounding and seismic measurements made by project GEORIFT. (Source: Tiercelin and Mondeguer, 1991.)

2.3.5 Water budget

The annual variation in the present water level of Lake Tanganyika is <1 m with long-term variations of about the same magnitude (Gillman, 1933; Edmond *et al.*, 1993). The Ruzizi River is the major contributor to both the water and salt budgets of the lake (Stoffers and Hecky, 1978). An estimation of the water budget of the lake is given by Edmond *et al.* (1993) on the basis of a combination of direct measurements of stream flow and estimates of the rate of precipitation and evaporation on the lake surface, as detailed in Table 2.2.

Table 2.2 Estimated water budget of Lake Tanganyika (Edmond *et al.*, 1993)

Water movement	**Flow rate** (km^3/year)
Outflow from Lake Kivu (controlled by hydroelectric dam)	3.2
Average inflow from the River Malagarasi	6.9
Inflow from ephemeral streams draining walls of the rift	8.1
Total discharge into Lake Tanganyika	**18.2**

The total discharge is equivalent to a layer 0.57 m deep over the lake surface. By comparison, the estimated precipitation on the lake (29 km^3/year) and evaporation from the lake (43 km^3/year) correspond to layers of 0.9 and 1.34 m, respectively. Thus the water excess to the Lukuga River averages 4.2 km^3/year, i.e. the present outflow is *ca* 9% of the total input, the remainder being removed by evaporation (Edmond *et al.*, 1993). Similar data given by Gasse *et al.* (1989) indicate that precipitation on the lake surface and surface run-off represent about 63 and 37% of the water input, respectively, and that 94% of the water loss is via evaporation.

2.3.6 Characteristics of the water column

Lake Tanganyika has a permanent thermal stratification, albeit a weak one. According to Beauchamp (1939) the thermocline extends to a depth of about 400 m and is strongly affected by seasonal variations in the strength of trade winds. Stoffers and Hecky (1978) report that the perennial thermocline is, however, at a depth of approximately 100 m. Surface water temperatures are reported to range from 23.3–29.5 °C (Hutchinson, 1975). Lowest surface temperatures are recorded in early August, whereas highest temperatures are characteristic of the March–April rainy season. At a depth of 40 m the temperature is reported to range from 26.3–29.5 °C. Below 200 m in depth, the mean temperature is 23.45 °C. It falls to no lower than 23.7 °C between 500 and 800 m and ranges from 23.32–23.35 °C below a depth of 1000 m (Hutchinson, 1975). Edmond *et al.* (1993) report that deep-water temperatures in the lake have been stable since 1939. Seasonal variations are confined to the top 200–250 m of the water column.

Beauchamp (1939) presented the first chemical profiles from Lake Tanganyika. He demonstrated that the water column is anoxic below about 100 m and that there are strong vertical gradients in silicate and phosphate across the thermocline. The latter may play an important role in promoting the flocculation of fine sediment particles settling in the lake waters (see later). According to Hutchinson (1975) the entire body of water below *ca* 200 m is richer in dissolved solids than the surface layers and is entirely devoid of oxygen. Little oxygen is present below 150 m in any season. Degens *et al.* (1971) report that the boundary between oxic and anoxic water masses occurs at a water depth between 80 and 140 m. Thus, Lake Tanganyika may be described as a meromictic lake, i.e. one in which part of the deep water is stabilized by dissolved substances. Hutchinson (1975) describes a meromictic lake as having a chemically stabilized zone of dead water at the bottom. In Lake Tanganyika, heating from the earth's interior (Botz and Stoffers, 1993; Stoffers and Botz, 1994) must cause slow circulation of the bottom waters. Furthermore, there is evidence that a sudden, heavy fall of cool rain may reduce the stability sufficiently to induce local mixing down to about 400 m into the otherwise stagnant water (Capart, 1952).

Implications of stratification for river inflow

The nature of the stratification of the water column, both thermal and chemical, is of critical importance to the mode of introduction of river-borne suspended sediments. Three basic types of inflow water movement can arise when a river enters a lake. These depend on the density difference between the lake water and river water which is governed by temperature and the concentrations of dissolved and suspended solids present. If the inflowing water density is less than that of the water at any level in the lake, the river water will tend to flow as a surface current (overflow) but usually becomes rapidly mixed by wind-generated turbulence. Conversely, if the river water is denser than any level in the lake, it will descend to and flow along the floor of the lake (underflow) as a well defined density current. Such currents may follow the courses of submerged valleys as referred to above. If the density difference is the result of temperature or dissolved solids the current is a true density current, whereas if the difference is due to a high concentration of suspended solids in the river water the current is best described as a turbidity current, its identity being lost when its load is deposited. True density currents lose their identity by diffusion and mixing with the surrounding water. Flow conditions transitional between overflow and underflow may occur, particularly when water enters stratified lakes, as a result of the density of the river water being equal to that at some depth below the surface (interflow).

Little is known of the modes of river inflow to Lake Tanganyika. However, it is reported that warm but saline waters enter from Lake Kivu via the Ruzizi River. These waters are denser than the surface waters of Lake Tanganyika, particularly during heavy floods (Tiercelin and Mondeguer, 1991), and move along the gentle slope of the shallow north end bringing heat, suspended mineral matter and oxygen into the monimolimnion (Hutchinson, 1975). In this way sediment by-passes the Ruzizi delta and accumulates on the platform and across the slope. Quartzo-feldspathic and micaceous sands characterize the platform, chiefly in the axes of the Petite Rusizi and Grande Rusizi channels. Small amounts of biogenic debris are present. The slope, from a depth of 40 m, is characterized by silts and clays particularly rich in gastropods, ostracodes, shell and fish fragments. The main sedimentation area appears to be associated with the Grande Rusizi channel as indicated by the wide lateral migration of sediment due to strong littoral currents (Tiercelin and Mondeguer, 1991).

The ultimate fate of suspended sediments when entering Lake Tanganyika will also be strongly influenced by the following factors: the mineralogy of the suspended particles, the chemistry of the receiving waters and the lake hydrodynamics. The water chemistry will control whether or not particle aggregation, or flocculation, of specific clay and other mineral species will take place. The hydrodynamics of Lake Tanganyika have been described in detail by Coulter and Spigel (1991). Of potential importance to sedimentation processes is the occurrence of internal seiches, oscillations of the thermocline. Even in the presence of winds which may act to damp the oscillation, such seiches can persist in the lake for periods of 6–7 months (Coulter and Spigel, 1991). Internal seiches create circulation patterns dominated by periodically reversing horizontal currents in nodal positions, and vertical motion at antinodes (McManus and Duck, 1988). Such seiche-generated currents can play a significant role in the promotion of collisions between suspended sediment particles possibly leading to orthokinetic (i.e. shear-induced) flocculation. Should flocculation, creating dense aggregates of primary particles, be important, this will exert a strong control on the dispersal of sediments to the offshore zones of the lake. Moreover, turbulence of the water column may be sufficient to cause the break up of flocs and will therefore play an equally important role in determining sedimentation patterns. Thus in addition to estimates of the sediment loads transported to the lake by influent rivers, information on the mineralogy of the sediments determined by X-ray diffraction, and on the chemistry/flocculation potential of the receptor waters, must be acquired. Whilst there is limited information from cores on the mineralogy of sediment deposited in the lake (see Figure 2.3), the apparent lack of data on influent sediment mineralogy is seen as a significant gap in our knowledge of the lake, a fact which is confirmed by discussions with specialists on the area (P. Coveliers, Tauw Milieu, Antwerp, personal communication).

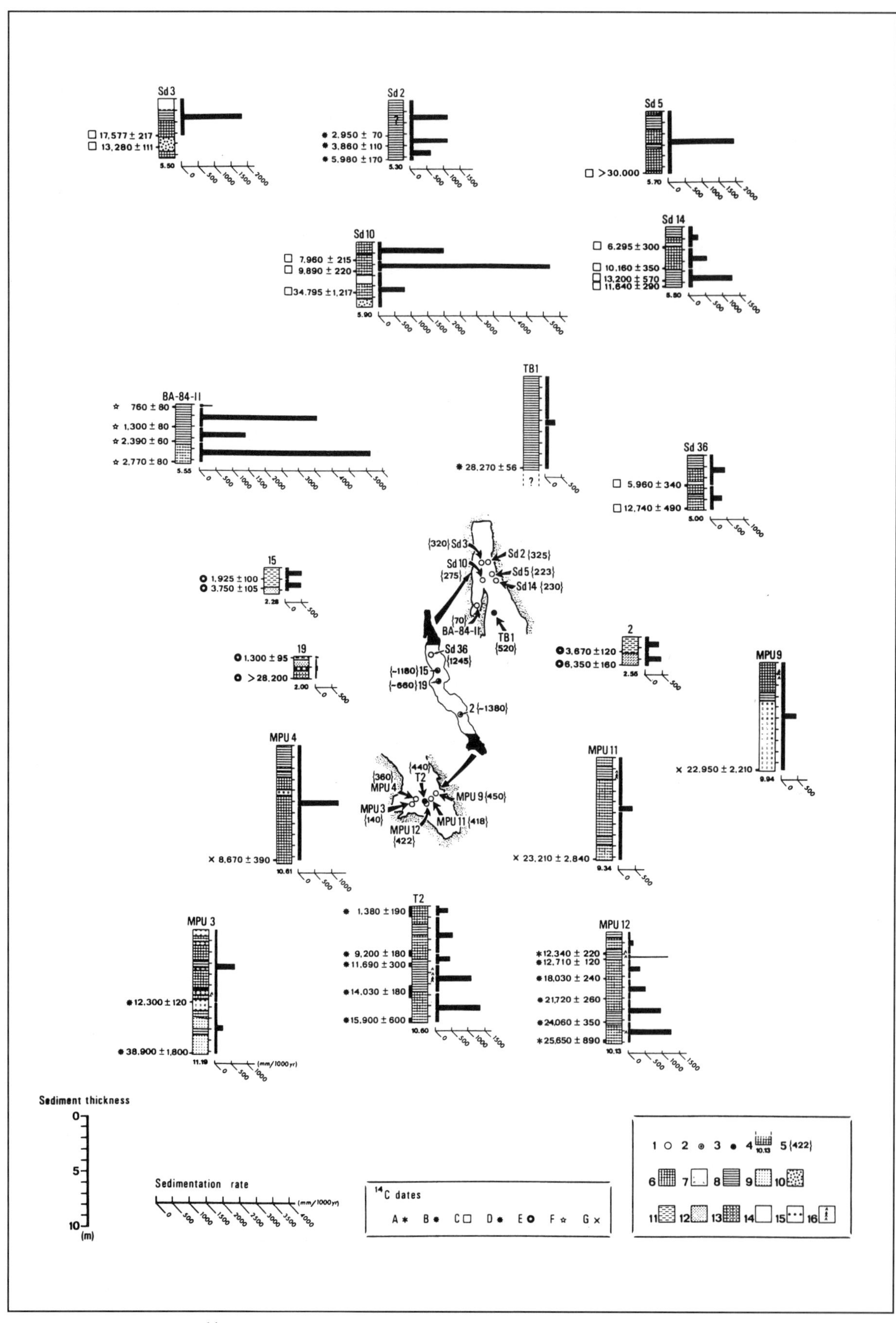

Figure 2.3 A review of ^{14}C dating on Lake Tanganyika sediments and corresponding sedimentation rates. 1–3, coring sites; 4, maximum length of cores (m); 5, water depth at coring site (metres below lake level); 6, homogeneous mud; 7, flaky mud; 8, laminated mud; 9, silt and silty clay; 10, coarse sand and gravel; 11 carbonate-rich facies; 12, kaolinite-rich facies; 13, smectite-rich facies; 14, 'Argile Beige', nontronite layer; 16, pyroclastics. (Source: Tiercelin and Mondeguer, 1991.)

2.3.7 Water colour and Secchi disc observation

The Secchi disc, the simplest of all oceanographic/limnological instruments, provides crude estimates of the turbidity of a water column in relation to the depth at which a white disc ceases to be visible below the water surface. Capart (1952) reported that the open water of Lake Tanganyika had Secchi disc visibilities in excess of 22 m. This isolated piece of information suggests extremely good visibility through and clarity of the water column, indicating low concentrations of suspended sediments and/or plankton. Coulter (1968), however, reported a maximum Secchi disc visibility of 17.5 m during 1960–62 at the extreme southern end of the lake. He found that Secchi depths were quite variable over the annual cycle and were apparently correlated with plankton abundance. Full details are given by Hecky (1991).

The colour of a lake refers to the colour of the light as it emerges from the lake surface. There are several empirical scales for recording colour, one of which is termed the Forel–Ule colour scale after its developers. It consists of a series of tubes containing solutions varying from blue through green to yellow and brown. The colour may be compared with that of a Secchi disc before it disappears, the standards being on a white background. The scale comprises XXI divisions: I, II (blue), III, IV (greenish blue), V, VI, VII (bluish green), VIII, IX, X (green), XI, XII, XIII, XIV, XV (greenish yellow), XVI, XVII, XVIII, XIX (yellow) and XX, XXI (brown). Lakes of colour I are extraordinarily rare (Hutchinson, 1975). Lake Tanganyika, however, is reported to have a colour of II in the "open part of the lake" and colours of III–IV "in rocky bays" (Hutchinson, 1975). These very limited data suggest that, at least at the time and location of the observations, suspended sediments were not being transported out to the central parts of the lake but were remaining in the nearshore waters from which they would settle to the bed. Whether or not flocculation processes are operative in Lake Tanganyika remains to be explored.

2.3.8 Sediment pollution and its consequences

Recent studies by Cohen *et al.* (1993) have focused on the impact of increased sediment input from rivers on the biodiversity of Lake Tanganyika (section 2.2). Forest clearing by major, uncontrolled fires has been followed by conversion of the former forest lands to grazing or for use in subsistence agriculture. Such clearing leads to rapid headward erosion, stream incision and gully erosion. The problem is not restricted to Lake Tanganyika (see Bruijnzeel, 1990; Cohen, 1991 for details). For example, almost 1000 km to the east, in Tanzania, deforestation and land degradation, accelerating in recent decades, have been identified as the cause of progressively more severe and destructive floods in the Lake Babati catchment in 1964, 1979 and 1990 (Sandström, 1995).

Bizimana and Duchafour (1991) have estimated soil erosion rates in the deforested, steeply sloping and intensively cultivated Nthangwa River basin of north Burundi to be between 20 and 100 t/ha/year. Nearly all of this sediment is discharged into Lake Tanganyika. A consequence of accelerated sediment discharge is the progradation of large deltas. For example, the delta of the Ruzizi River (see above) appears to have undergone an order of magnitude increase in its rate of outbuilding subsequent to major deforestation in its drainage area.

Cohen (1991) studied diversity data for three well documented groups of organisms in the coastal waters of Lake Tanganyika: ostracodes, fish and diatoms. Their sample coverage was extensive with only the central and southern coasts of Republic of Congo not represented in the study. In the virtual absence of directly measured suspended sediment concentration data for the influent rivers, sediment loading was measured qualitatively on the basis of the size of sediment plumes entering the lake and on the proportion of deforestation and channel incision evident in Landsat images and aerial photographs. Using these criteria Cohen (1991) classified all sampling sites into one of three categories of 'sedimentation disturbance': lightly, moderately or highly disturbed.

Ostracodes were found to be significantly less diverse in highly disturbed sites than in less disturbed ones for both hard and soft substrate littoral environments, with reductions in species richness ranging from 40–62%. Species richness patterns for profundal ostracodes show smaller differences between low and high disturbance sites (7–32%). The data for fish show a similar pattern to those for ostracodes, although the data are too limited to be analysed statistically. Cohen (1991) also report that unpublished studies of fish diversity at additional low and moderate disturbance sites in northern Republic of Congo and high disturbance sites in northern Burundi confirm their observed patterns. Diatoms showed only minor and statistically insignificant reductions in species diversity between low and high disturbance sites (15–20%). It was suggested that ostracodes and fish are more affected by sedimentation because they are mostly endemic and require clear-water habitats. The benthic diatom species in Lake Tanganyika, however, are largely cosmopolitan and in many cases also inhabit the turbid waters of influent rivers such as the Ruzizi.

In order to confirm whether the low species diversity at highly disturbed sites is due to increasing sedimentation rates through time, it will be necessary to examine sediment cores and the fossil records they contain. Cohen *et al.* (1993c) aim to use ^{210}Pb dating to evaluate changes in sedimentation rates; pollen analysis as a means of assessing the timing of the introduction of agricultural species; and ostracode and diatom fossils to monitor changes in biodiversity through time. This approach is considered to be important. Whilst it is relatively easy to assess modern rates of suspended sediment influx to the lake from a given river, previous data on transported sediment loads are seldom (if at all) available. Therefore the sediment record as preserved in the lake must be examined, in the form of cores, to estimate sediment loads in historical times by these indirect means. In this way, 'before' and 'present' data on suspended sediment influx to the lake may be compared at specific sites.

According to the findings of Cohen *et al.* (1993), the propensity of a drainage basin to severe erosion is primarily a function of its size and its soil/geological characteristics. The areas at lowest risk of increased sediment loss are those with small drainage basin size, and with bedrock which is both generally resistant to erosion and generates few fine particles during erosion (e.g. quartz sandstones). Such basins characterize the Tanzanian coast of the lake, both to the north and immediately south of Kigoma. By contrast, the littoral regions most vulnerable to sediment pollution are those fed by large drainage systems. These include most of the Burundi coast (already deforested) and much of the Zambian coast (still extensively forested). The vulnerability of such areas is still greater where only small delta plains are developed, thus minimizing the subaerial entrapment of sediment particles.

2.3.9 Sedimentation rates

Limited data on sedimentation rates, as determined from ^{14}C-dated cores, are given by Tiercelin and Mondeguer (1991). They present data from various sources incorporating seven cores from the northern part, two from the north-central part, one from the south-central part and six from the southern part of the lake (Figure 2.3). In general terms the northern part of the lake has experienced higher rates of sedimentation (maximum rate *ca* 4700 mm/1000 years) than the southern part (maximum rate <1500 mm/1000 years) whilst the cores from the central areas indicate still lower sedimentation rates (maximum rate <500 mm/1000 years). This supports the thesis that the Ruzizi River is a dominant source of sediments, despite the fact that the upstream Lake Kivu will be trapping materials in transit through the drainage network, and that the sediments delivered to the lake by the Malagarasi River appear to become largely trapped in its delta system. Two of the cores from the southern end of the lake indicate a decrease in sedimentation rate with time over the past 15 900±600 to 25 650±890 years. In the central parts of the lake, sedimentation rates appear to have remained approximately constant over the past 3000–6000 years. The pattern of sedimentation in the northern part of the lake appears to be the most complicated through time (Figure 2.3), individual cores indicating both increasing and decreasing sedimentation rates. Overall, highest sedimentation rates appear to have occurred over the past 3000 years in the relatively restricted waters of Burton Bay, supporting the suggestion above that fine particles are not generally transported far out into the lake. It should, however, be emphasized that the values for sedimentation rates suggested above take no account of the nature and origin of the materials accumulating on the bed, i.e. the relative quantities of minerogenic debris introduced by influent rivers and the biogenic debris (e.g. diatom tests) derived from within the lake itself.

It must be emphasized that the data collated by Tiercelin and Mondeguer (1991) are limited in their value. Cores specifically from the littoral zones of the lake should be collected and a greater resolution of dating over the past few decades is required. It is suggested that ^{137}Cs profiling will prove potentially important. ^{137}Cs profiles are now being widely used to assess recent rates of sediment accumulation in lakes (e.g. Walling and He, 1993). The basis for using ^{137}Cs is that radiocaesium is rapidly and strongly bound to fine sediment particles and its distribution in the sediment profile directly reflects the chronology of sediment deposition. The first appearance of this radionuclide can be dated to the early 1950s, and the vertical distributions of ^{137}Cs in the sediment profile can be related to the known record of fallout for the subsequent period (Payne, 1985). The peak fallout levels that occurred in 1963 have often been used to date sediments deposited at that time. Thus the ^{137}Cs method can provide important information on sedimentation rates during the last four to five decades, the time period during which deforestation and consequent land degradation have accelerated in the various catchments draining into Lake Tanganyika.

From the literature searches undertaken and discussions with various experts, it appears that there is no information on contemporary rates of sedimentation in Lake Tanganyika as acquired from sediment traps suspended above the lake bottom. Such traps could be deployed to monitor the spatial and temporal patterns of sedimentation at sites off the mouths of influent rivers and at other appropriate locations in the lake. This lack of information is identified as a further gap in our knowledge of sedimentation processes in the lake and it is recommended that sediment traps should be installed at an early stage of any field programme in order to monitor sedimentation rates over as long a time period as possible.

2.3.10 Drainage into Lake Tanganyika

The behaviour of the rivers reaching the shores of the lake is strongly dependent upon the rainfall, the proportion of precipitation retained by the vegetation or recycled into the atmosphere, and the nature of the underlying soils. In addition to the three principal influent streams, Ruzizi, Malagarasi and Lugufu, which drain large catchments, there are scores of minor, often ephemeral inflows from small catchments. The sole effluent stream flows from the lake at Kalemie as the Lukuga, to join the Lualaba tributary of the Zaire River, whence to the Atlantic.

Each of the bordering countries maintains a series of rainfall, climatological and hydrological stations. In some cases the rainfall records extend from before the start of this century, others provide short-term snapshots of conditions and at some sites the recording has ceased for various reasons. Many of the records, linked within the FRIEND (Flow Regimes from International Experimental and Network Data) programme, are accessible within the offices of the relevant hydrometric agencies. Information on the locations of gauging stations and the years for which records are available in catchments in both Tanzania and Zambia has been collected.

Rainfall

Rainfall is the fundamental control on river flow, and has been measured for longer than any of the other variables. It is still the most widely determined characteristic of the systems concerned. In Tanzania as a whole, for example, there are 348 rainfall stations and 89 climatological stations, but only 35 hydrometric stations (Mihayo, 1993). Many of the rainfall gauging stations are voluntarily operated and the records are often discontinuous. Nevertheless most of the data are reported to the Directorate of Meteorology on a monthly basis, providing a good level of information on 24-h total rainfalls.

The annual precipitation over most of the 151 900 km^2 Tanzanian catchments draining into Lake Tanganyika varies between 800 and 1200 mm/year. The FRIEND hydrometric maps of mean annual rainfall confirm that the bulk of the Malagarasi system receives between 800 and 1000 mm but that the headwaters are drier, with between 600 and 800 mm. In the extreme north beside the Burundi border, rainfalls of 1000–1400 mm/year are recorded, and in the extreme south a zone receiving over 1000 mm of precipitation reaches more than 25% of the lake shoreline. The rain falls heavily during the period from October until April, with very little during the other months of the year (Figure 2.4, from Coulter and Spigel, 1991).

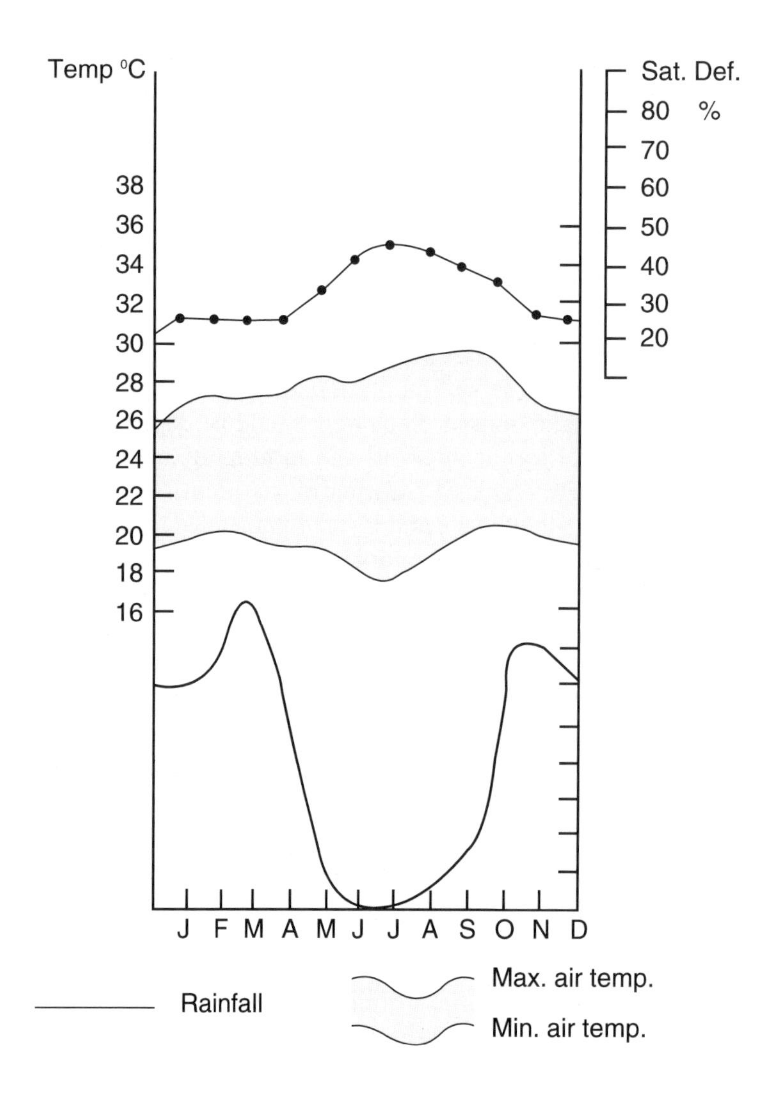

Figure 2.4 Average monthly rainfall, maximum and minimum air temperature and saturation deficit at the shore of Lake Tanganyika at Kigoma, Tanzania. Rainfall is averaged over 18 years, and temperature and saturation deficit over 12 years within the period 1930–56. (Source: Coulter and Spigel, 1991.)

In a contribution to the DUSER Project (Dar es Salaam/Uppsala Universities Soil Erosion Research Project, examining soil erosion in Tanzania during the period 1968–72), Jackson (1972) provided a mean annual rainfall map for the country (Figure 2.5) from which it is possible to recognize that the headwaters of the Malagarasi stretch as far east as the 1000 mm isohyet. The Lake Tanganyika catchments are separated from the areas of semi-arid terrain in the centre of the country. Although there was no rain in July and very little in April, in both the January and October monthly analyses rainfall intensities exceeded 25 mm/day along the south-eastern shores of the lake (Figure 2.6).

Analysis of records from 1880–1980 enabled Lema (1990) to conclude that there was no evidence of climatic change having occurred in East Africa during the last century, a finding supporting the conclusions of Rodhe and Virji (1976) who examined records from 1920–60, and those of Ngama (1992).

It will be possible to obtain primary rainfall data from each of the countries directly but, as a systematic analysis of the records would not have been possible within the time-scale of this study, this was not attempted. Analysis of data from small isolated catchments around the lake shore would reveal much about the arrival of water within systems from which it will pass rapidly into the lake waters.

The rain falling on a catchment may run off into the rivers or may soak into the soils, where it is either used to support plants or soaks downwards to join the subsurface groundwater; the remainder may become recycled by means of evaporation or returned to the atmosphere in evapotranspiration.

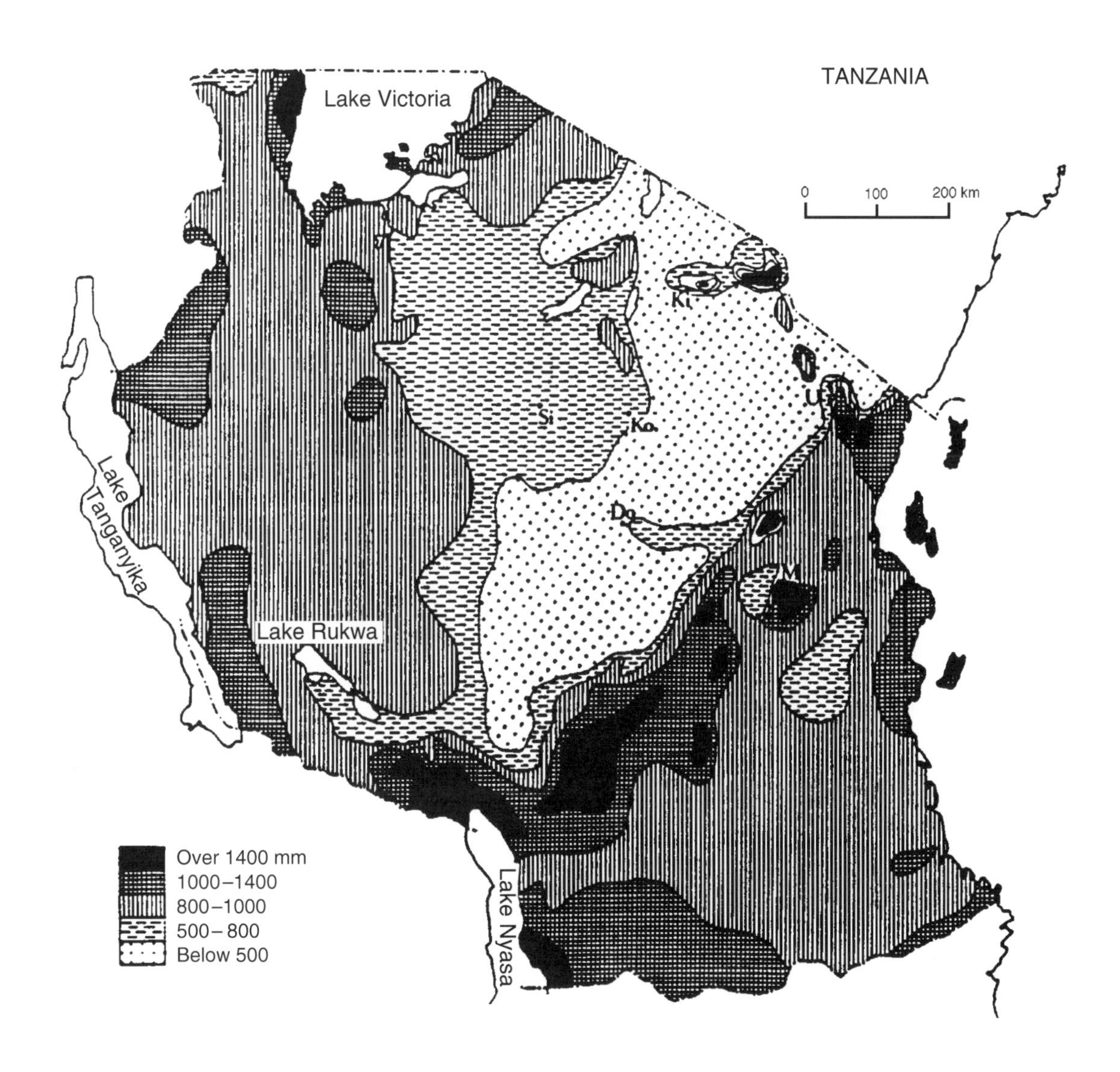

Figure 2.5 Map of mean annual rainfall (after Jackson, 1972) and field areas of DUSER catchment field studies. (Source: Rapp *et al.*, 1972a.)

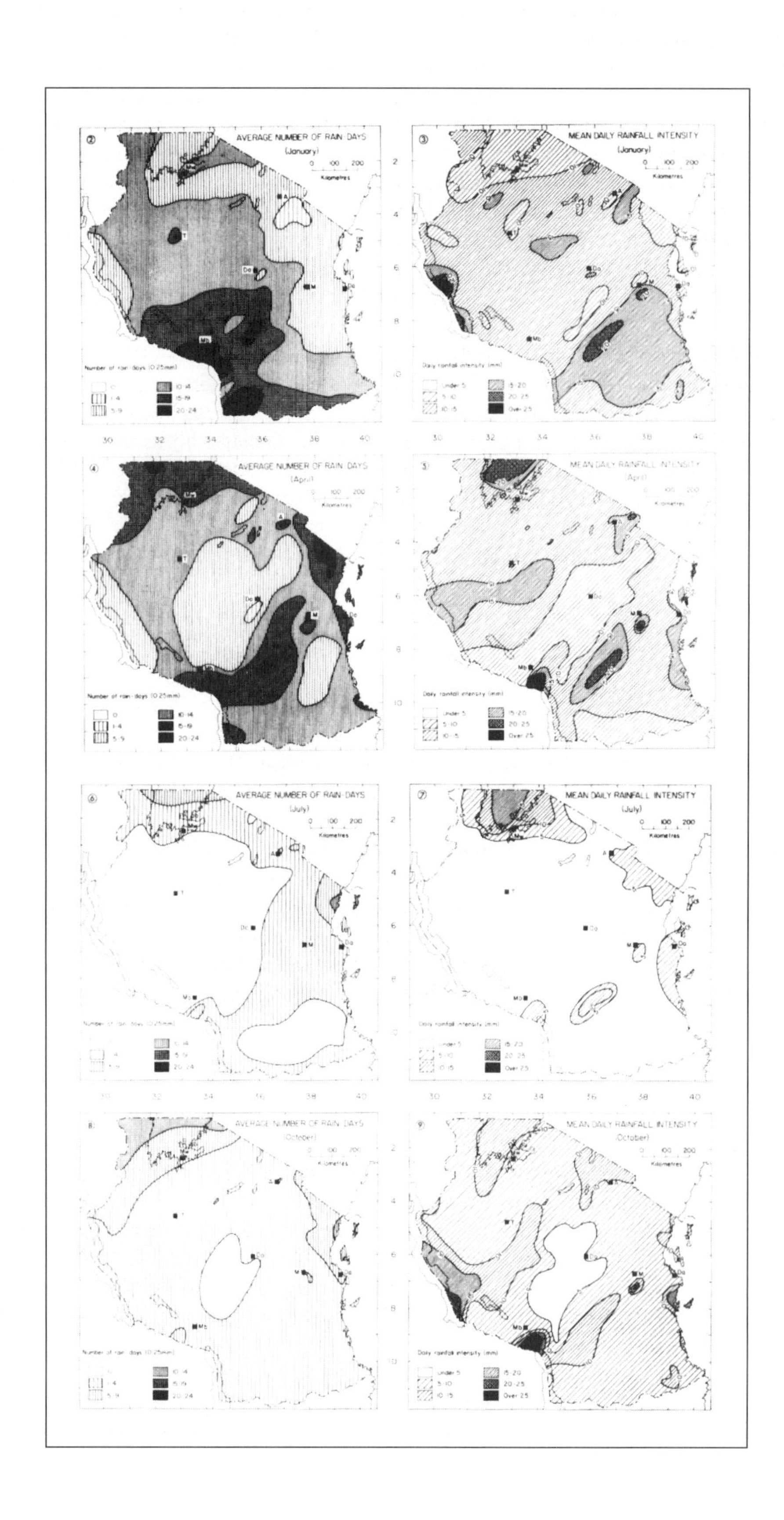

Figure 2.6 Spatial variation of average number of rain days and mean daily rainfall intensity over Tanzania. (Source: Jackson, 1972.)

River flow

River gauging stations are present on all of the major influent rivers to Lake Tanganyika. The locations of the principal gauging stations in Tanzania are indicated in Figure 2.7 (from DANIDA, 1995, VIII-3), and a list of Hydrometric Stations in the Lake Tanganyika Basin (DANIDA, 1995) is given in Table 2.3. Whereas for the major gauging stations the records provide both stage height and water discharge, often coupled with additional information such as suspended matter concentrations and chemical analyses, for many sites of lesser quality the records consist solely of stage level readings with poor control of the calibration to convert the figures into discharges. The details of gauging stations in the rivers of Tanzania and Zambia are summarized in the FRIEND Master Register (UNESCO, 1994). A detailed analysis of the characteristics of the basins flowing into Lake Tanganyika is given in Volume 2 of the DANIDA report (1995). Although figures for discharges of various rivers are quoted, it should be noted that the DANIDA report casts considerable doubt on the reliability of the hydrometric data from Tanzania, pointing out that at many gauging stations the meters and gauging cableways are missing, and also that rating curves are rarely checked due to a lack of current meters.

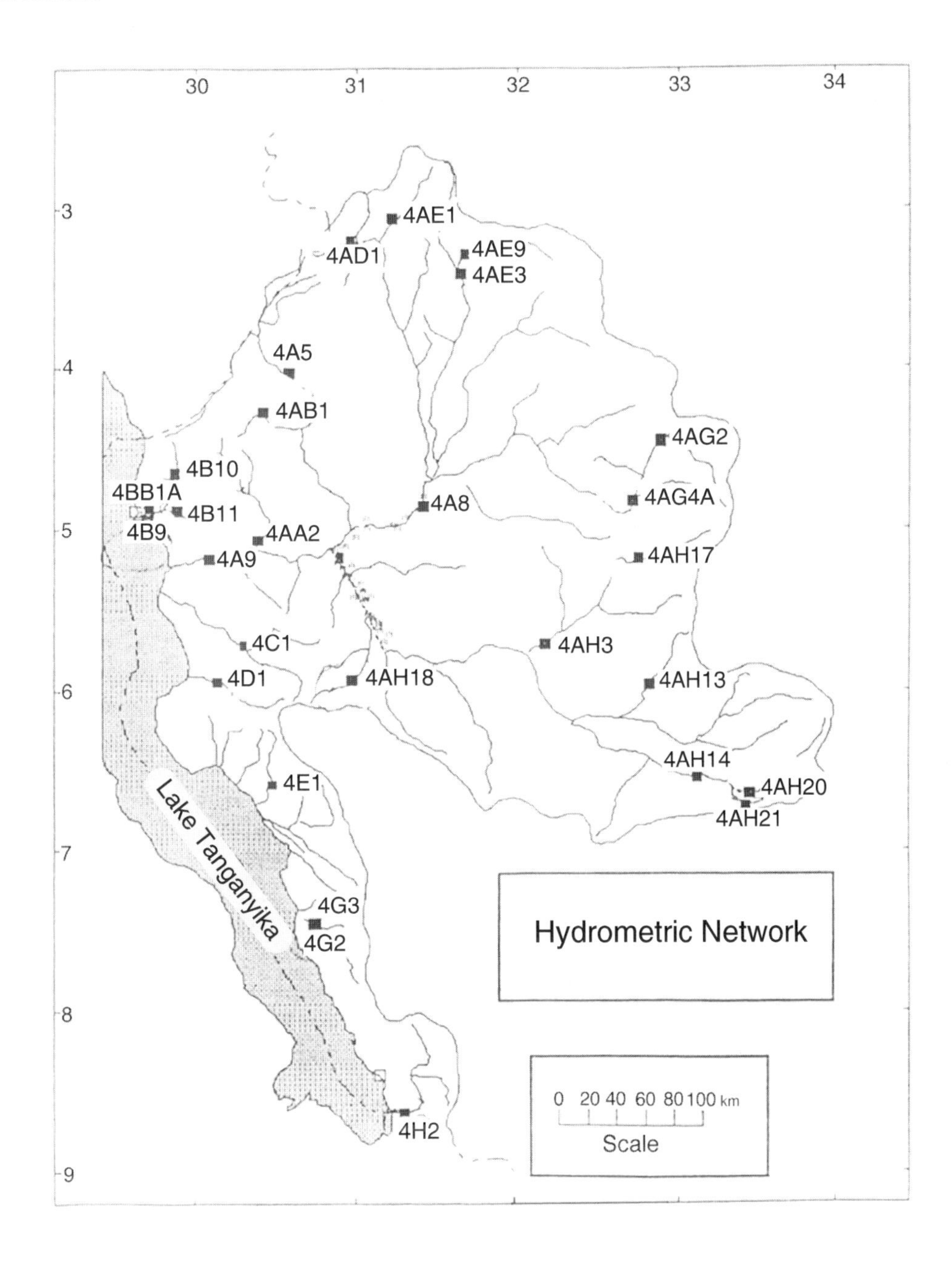

Figure 2.7 Hydrometric Network for Tanzania. For site names see Table 1.3. (Source: DANIDA, 1995.)

Table 2.3 List of Hydrometric stations — Lake Tanganyika Basin (DANIDA, 1995)

Code	Type	River name	Site name	Location Lat		Long		Area km^2	Start Date (End 1992)		Status at
4/1	W	Lake Tanganyika	Mpulungu	8	7	31	6	-I	APR	1957	
4/2	W	Lake Tanganyika	Kigoma	4	53	29	37	-I	DEC	1954	
4/3	W	Lake Tanganyika	Albertville	-4	54	29	15	-I	JUL	1959	
4A3	L	Igombe Dam	Spillway	-I	-I	-I	-I	-I		-I	
4A4	Q	Mwiruzi	Kanyoni	3	13	30	47	-I	JAN	1976	
4A5	S	Malagarasi	Taragi Bridge	4	2	30	35	8525	NOV	1961	
4A9	W	Malagarasi	U/S Nbelagule	5	11	30	5	123 300	APR	1975	
4AA2	Q	Ruchungi	Paga (Winza)	5	4	30	23	2825	OCT	1974	
4ABIA	Q	Makere DS	Makere Pump House	4	17	30	25	550	OCT	1974	
4ADI	Q	Ruiti	Nyakusange	3	12	30	58	-I	NOV	1974	
4AD2	Q	Myoosi	Kanyoni	3	12	31	1	3350	DEC	1975	
4AEI	Q	Kahogo	Nyakanazi	3	4	31	13	1050	APR	1976	
4AE3	S	Nikonga	Msasa	3	9	31	39	4200	NOV	1974	c1987
4AE4	S	Hikonga	Nyatakala	3	24	31	38	3800	APR	1976	c1979
4AG2	S	Igombe	Pala	4	22	32	53	1800	AUG	1974	
4AG4	Q	Nzubuka	Nzubuka Railway Bridge	4	48	32	49	1400	AUG	1974	
4AG4A	Q	Igombe	Igombe Idge	4	50	32	43	418	OCT	1976	
4AH2	Q	Mtambo	Nsan (Ugala)	5	59	31	11	800	OCT	1977	c1982
4AH3	Q	Ala	saw Mill	5	43	32	10	8848	SEP	1975	c1977
4AH7	Q	Ala	Pangale	5	11	32	45	1520	DEC	1974	
4AH13	Q	Ulua	Ngoyua	5	58	32	49	1740	SEP	1974	
4AH14	S	Nkululu	Nkululu	6	33	33	7	3800	OCT	1973	s1985
4AH17	Q	Mungo	Miungu	6	40	33	42	-I	JAN	1977	s1984
4AH18	Q	Mwalezi	Uvina Road Bridge	5	56	30	58	360	JUN	1977	
4AH20	Q	Nkululu	Magawe	6	39	33	27	1525	SEP	1994	s1984
4AH21	Q	Miungu	Magawe	6	40	33	25	700	SEP	1975	s1982
4B9A	Q	Luiche DS	Kigoma Simbord	4	54	29	42	2050	OCT	1977	
4B10	Q	Luiche	Jimbi	4	39	29	52	750	NOV	1974	
4BII	Q	Mkuti	Mkuti	4	53	29	53	143	FEB	1976	
4BBIA	Q	Kaseke	Kigoma-Simbord Road	4	53	29	41	-I	NOV	1960	
4DI	Q	Luegele	Lubalisi	5	53	30	1	1450	MAY	1975	
4FI	W	Msenguse	Ikola	5	42	30	25	-I	AUG	1976	c1980
4G2	Q	Luamfi	Masolo	7	28	30	44	533	JUL	1975	
4G2A	Q	Luamfi	Masolo	7	28	30	44	531	JUL	1975	
4G3	Q	Samba	Masolo	7	28	30	44	353	JUL	1975	
4HI	Q	Kalambo	Kalambo Falls	8	36	31	11	3120	JAN	1975	

The Malagarasi catchment is the largest entering Lake Tanganyika. Together with its many tributaries it drains an area of 130 000 km^2 which stretches from the Burundi and Tanzania mountainous border country through areas of low relief including the extensive Malagarasi–Moyowosi swamplands before flowing through reaches dominated by falls and rapids as it crosses the Misito escarpment. The principal tributaries are the Moyowosi and Igombe, which, like the Malagarasi, flow into the seasonal Lake Nyamagoma, the Ugalla and Walla which likewise create seasonal lakes (Sagara and Ugalla), and the right bank Ruchugi which has its confluence at Uvinza.

A substantial network of gauges exists within the Malagarasi catchment, with most of the main tributaries having their own gauges. A total of 12 of the 18 gauging stations have automatic water level recorders. The DANIDA *Rapid Water Resources Assessment* (1995) lists a further five gauging sites on the Luiche, a significant river which flows into the lake by way of a substantial delta near Kigoma.

Several other Tanzanian gauged rivers enter Lake Tanganyika. In a southward sequence these are the Luegele, Lufugu, Luamfi and Kalambo. Mean flows quoted for some of these gauges provide an indication of the relative importance of the streams (Table 2.4).

Table 2.4 Mean flows for some of the gauged rivers entering Lake Tanganyika

River	Flow rate (m^3/s)
Malagarasi at Mberagule	156.6
Luiche at Kigoma Simbord	19.4
Luegele at Lubalisi	9.7
Lugufu at Katentye	42.4

The gauged rivers of the Republic of Zambia which flow north into Lake Tanganyika are listed in the *Hydrological Year Books* of the Ministry of Agriculture and Water Development. The lowest gauging stations on the principal lake influents are at Kambole on the Lunzua River and at Keso Falls on the Lufubu River. The mean flows listed show discharges of 4 and 14.9 m^3/s, respectively.

Discussions with P. Coveliers of Tauw Milieu, Antwerp have revealed sources of information on the inflows of rivers draining into Lake Tanganyika from Republic of Congo and Burundi, including the principal influent from the north, the Ruzizi River. These data include material held by the Royal Belgian Museum of Central Africa in Brussels, curated by Professor Jean Klerckx, the Director of the Geology Department of the Museum, and information gathered by various scientists during specific research projects. A full inventory of this database has yet to be compiled but it is known to include information not only on the Ruzizi River but also on at least 12 rivers of Republic of Congo (P. Coveliers, personal communication). The latter were studied during 1992–93 when monthly sampling of water discharge, turbidity, pH, electrical conductivity and dissolved oxygen concentration were carried out at stations located *ca* 100 m upstream from the lake.

Water escapes from Lake Tanganyika by way of a single outlet near Kalemie, where a gauging station is listed as measuring discharges. A flow of 90 m^3/s was quoted by Rodier (1983) for the effluent Lukuga River measured at Kalemie.

Run-off

As indicated above, water derived from the rainfall reaching the land surface may flow into rivers, may soak into the soils or may return to the atmosphere by evaporation or through plants as evapotranspiration. The quantity of rain falling on a catchment does not necessarily provide a good indication of the quantity of water which will escape into the rivers, whence to the lake. Likewise, retention of the river waters within swamplands increases the time and thereby opportunity for evaporation to occur, still further reducing the quantity of water in transit.

According to DANIDA (1995), the smaller catchments successfully deliver a large proportion of their water budget (between 22.1 and 33.6%) to Lake Tanganyika, whereas the larger rivers of the Malagarasi system, all of which flow through the Moyowosi or Ugalla swamps, carry only between 2.6 and 4.1% to gauging stations below the swamps.

Studies reported by Sandström (1995) in the catchment of Lake Babati, some 300 km east of the Malagarasi headwaters, suggest that in the upland areas the run-off accounted for 3.5–26% of the annual rainfall, depending on the vegetation cover and the soils present. Forested areas yielded a 2% run-off, whereas in deforested areas the run-off rose to 10%. From analysis of air photographs, Newman and Rönnberg (1992) demonstrated that land use had changed drastically in the area, so that between 1960 and 1990 the area under cultivation had increased from 23 to 58%, with a corresponding decrease in woodland and bushland from 73 to 38%. Sandström (1995) concluded that over the period from 1940 the rainfall run-off from the one catchment had doubled by 1990. There is no reason to suppose that the population pressures in the headwater regions of the Malagarasi system are any less, so it is likely that there will have been an increase in the proportion of rainfall entering the rivers in such areas. However, as the majority of the headwater streams pass through the swamplands, much of this change will have been absorbed before emerging lakeward of the retention area.

Sediment yield

It is now more than a decade since, in an abstract to a review of sediment yields of African rivers Walling (1984) wrote "Although measurement of sediment transport by African rivers can be traced back to the nineteenth century relatively little is known about sediment yields within the continent and many uncertainties exist". He noted the generalized figures of 50–100 t/km^2/year provided by Strakhov (1967) for catchments in central and eastern Africa, and the much larger 600–2000 t/km^2/year derived by Fournier (1960) using relationships between sediment yield, basin relief and precipitation. Walling (1984) provided a similarly generalized map suggesting that the catchments draining into Lake Tanganyika should yield suspended sediment at a rate of 100–1000 t/km^2/year.

In a more recent review of the yield of sediment as a function of drainage area in differing geological settings, Milliman and Syvitsky (1992) suggested that in areas of thermal extension of the crust, as in the rift valley setting, sediment loss varied with the size of the drainage basin. Yields were shown to decrease from 110 t/km^2/year in 500 km^2 basins to 30 t/km^2/year in very large basins of several million km^2 (Figure 2.8), stressing the important role played by small rivers in the denudation of land surfaces. This is of great significance to the Lake Tanganyika system, for, in contrast to the few large rivers which drain into it, many small catchments lead directly into the lake without intervening swamplands.

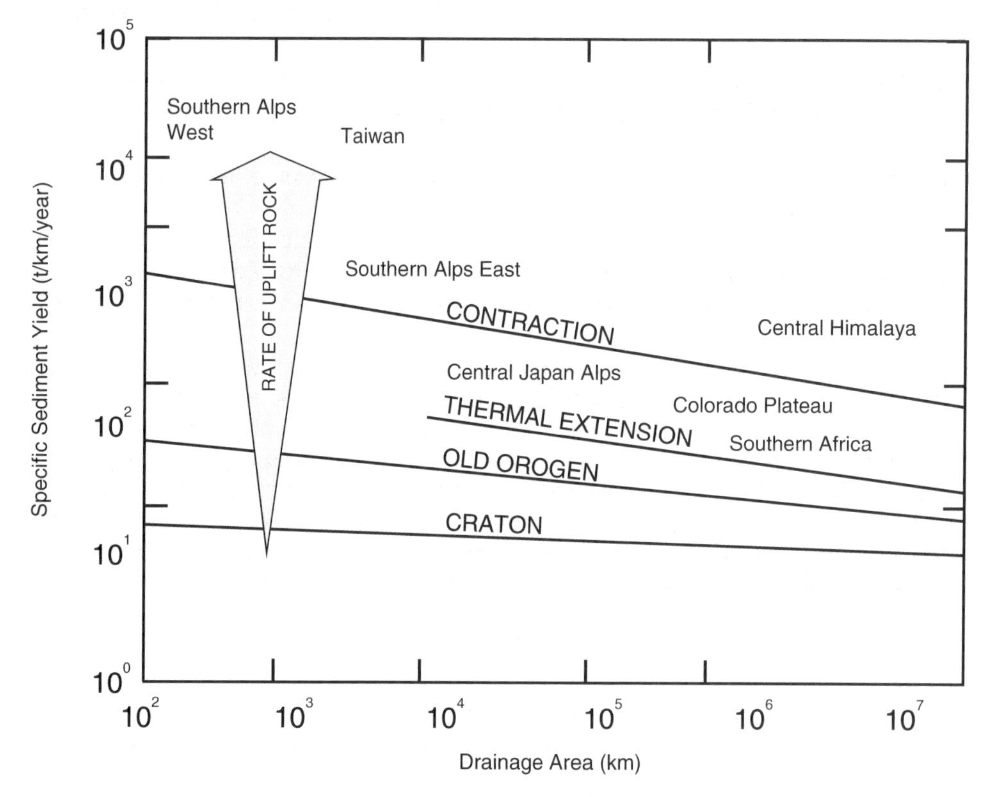

Figure 2.8 Summary of relationships between sediment yield and basin area in a range of geological settings. The line marked 'thermal extension' is that appropriate for the Lake Tanganyika area. (Source: Milliman and Syvitsky, 1992.)

These figures are marginally greater than those of El-Swaify *et al.* (1982) who concluded that average African soil-loss rates were approximately 72 t/km^2/year, of which about 32% was in solution.

While there have been many studies on field or plot scale in Tanzania, there has been no systematic approach to the determination of sediment transport within the river systems. According to DANIDA (1995) sediment sampling is undertaken at about 30% of the gauging stations, but although sediment–discharge relationships are known for some stations there appears to be no established programme to use the resultant data to estimate sediment yield or soil losses from the catchments.

The findings of the DUSER project, carried out during the period 1968–72, are recorded in 15 papers in a special issue of *Geografiska Annaler* (1972). Many of the studies took place outside the Lake

Tanganyika catchments, but they nevertheless provide an indication of factors likely to be important for the lake study, namely slope gradient, length, rainfall duration and intensity, soil characteristics, and particularly vegetation cover. Not surprisingly these are the main factors identified in the theoretical universal soil loss equation (Wischmeier and Smith, 1978).

Slopes of gradients above 30° are highly susceptible to landslide development and severe soil losses (Rapp *et al.*, 1972b). Soil losses increase with rainfall intensity and duration of the rainfall, a pattern confirmed in simulated rainfall studies in Kenya by Snelder and Bryan (1995). These researchers also showed that vegetation cover exerts an important influence, with greatest sensitivity to storm run-off being exhibited by areas with 25–55% plant cover. In areas exceeding the critical 55% vegetation cover, sediment yields fell drastically, and they recommended that where possible the land surface should be returned to a condition bearing a 50% plant cover, in order to control soil erosion.

In the Shinyanga region, which includes the headwater areas of the Igombe River, Stocking (1984) demonstrated that erosion rates of 1.4, 10.5 and 22.4 t/ha/year could be detected in old grazed land, 20–30-year-old grazed land, and newly developed grazing lands, respectively. Losses of 19.6–61.6 t/ha/year were normal in uncultivated areas, and 36.5–70 t/ha/year in newly ploughed land. From studies during three rainy seasons Rapp *et al.* (1972b) show that there were negligible soil losses from forested lands, rather more from fallow grassland, but much more from newly ploughed land. As noted earlier, the presence of a good vegetation cover also serves to reduce water run-off, a conclusion supported by comparative studies in Zimbabwe, Uganda and Tanzania (Temple, 1972).

The influence of land use on sediment yield in nearby Kenya is well illustrated by Figure 2.9 (Dunne, 1979), which suggests that the density and type of vegetation exerts a major influence on soil loss. The soil losses have often been calculated from field plots, commonly 0.1 ha in area. It is not necessarily correct to multiply the figures by 100 to convert them into yields in t/km^2/year for comparison with catchment scale sediment yields, because there is no guarantee that all of the material will indeed be carried from the drainage area. It is quite normal for sediment transported from field plots to become redeposited within the catchment, where it enters a form of temporary storage. According to Walling (1984) possibly no more than 10% of the released sediment is swept from the catchment, even where there are no lakes or swamplands to intercept material in transit. Berry and Townshend (1972) considered that the delivery ratio lay between 10 and 20%, the larger figure being supported by Rapp (1977) on the basis of studies in five Tanzanian catchments. Data on sediment yields to Lake Tanganyika from rivers of Republic of Congo and Burundi are apparently limited.

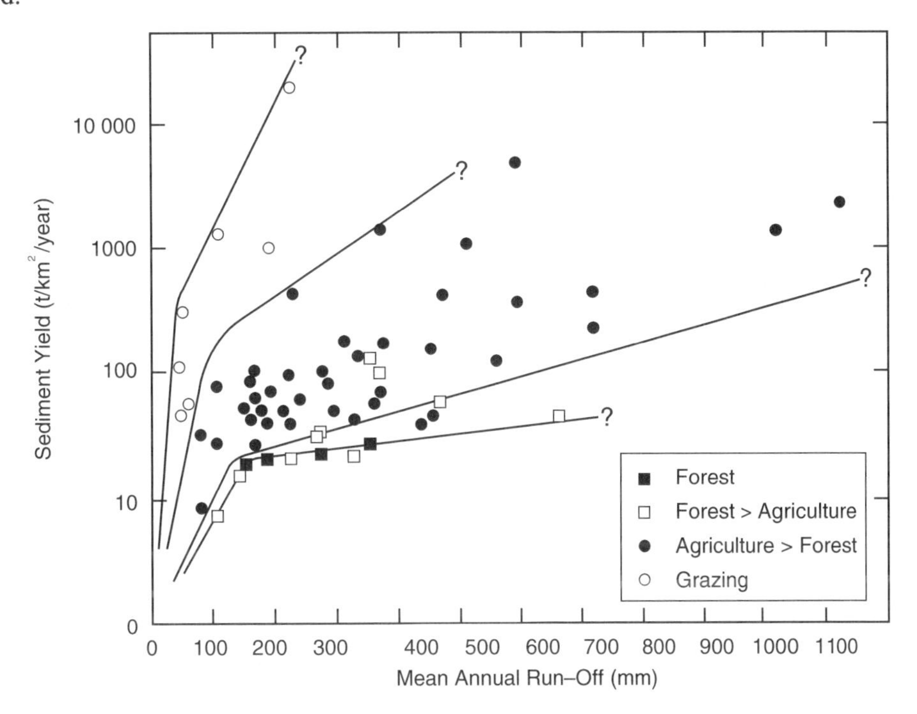

Figure 2.9 Mean annual sediment yield and mean annual run-off for catchments with indicated dominant land uses. (Source: Dunne, 1979.)

Soils

According to Baker (quoted by Berry and Townshend, 1972) "most Tanzanian soils are extremely erodible once the surface cover is removed". There are several versions of soil maps available. Perhaps the simplest, enabling general statements to be made, is that given by Masija (1993), who summarized the soils of the Lake Tanganyika catchment as principally loams or sandy loams, with sandy clay loams immediately beside the lake and occupying the Burundi borderlands. A substantial area of poor agricultural soils is marked between the mouths of the Malagarasi and the Luiche Rivers (Figure 2.10). Similar and more detailed soil survey maps are available for other catchments feeding into the lake.

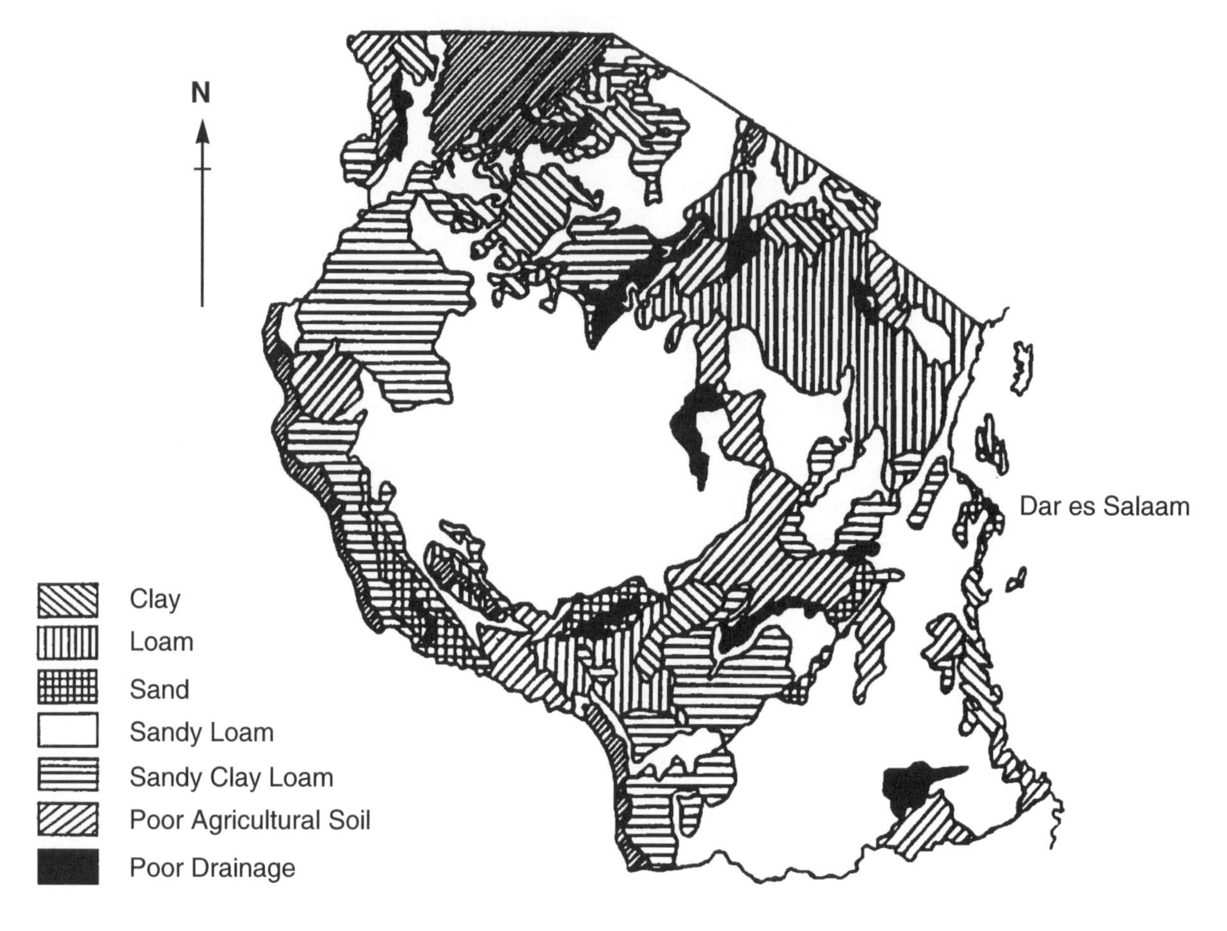

Figure 2.10 Soil textures of Tanzania. (Source: Masija, 1993.)

2.4 IMPACT OF SEDIMENT INPUT ON THE LIMNOLOGY OF LAKE TANGANYIKA

2.4.1 Introduction

By limnology we understand the functional relationships and productivity of fresh water communities as they are regulated by the dynamics of their physical, chemical and biotic environment, i.e. all relevant food-web interactions. This section will therefore include some aspects already mentioned in the other sections of the review.

2.4.2 Current level of scientific knowledge

Nutrients are generally considered to limit algal productivity in aquatic ecosystems when light and temperature are adequate for photosynthetic growth to proceed (as is the case in most parts of Lake

Tanganyika). No comprehensive studies on nutrient limitation have been done on Lake Tanganyika, but nutrients have often been evoked to explain differences in species composition and productivity (e.g. Hecky and Kling, 1987). Hecky and Kling (1981) specifically cite nutrient availability as being likely to control the seasonal succession of algal species in Lake Tanganyika, and discuss circumstantial evidence suggesting that either nitrogen or phosphorus (and silicon for diatoms) may at times limit algal growth over the annual cycle. Various authors have observed shifts in algal species with changing nitrogen/phosphorus ratios: low N/P ratios appear to favour nitrogen-fixing blue-green algae whereas high N/P ratios, achieved by controlling phosphorus input, cause a shift from a water bloom consisting of blue-green algae to one containing forms that are less objectionable.

The pelagic zone

A complete chemical screening of the Lake Tanganyika pelagic zone has not yet been undertaken. The spatio-temporal distribution of nutrients in the mixolimnion is still poorly defined, and even a complete annual cycle of epilimnetic nutrient concentrations has not yet been observed for any location on the lake.

Beauchamp (1939) reported the first chemical profiles of the lake. He showed that the water column is anoxic below *ca* 100 m and that there are strong vertical gradients in silicate and phosphate across the thermocline. He reported very high concentrations of ammonia in the anoxic zone. During 'L'exploration hydrobiologique du lac Tanganyika' (1946–47), J. Kufferath carried out biochemical studies (Kufferath, 1952). He presented oxygen, major ion and nutrient profiles which were confirmed later by Edmond (1993) and Edmond *et al.* (1995) They demonstrated that the oligotrophism of the lake was very special one, as it had such high productivity.

The most comprehensive studies of the geology and geochemistry of the lake reported until the study of Edmond *et al.* (1993) were those by Degens *et al.* (1971) and Hecky and Degens (1973). They presented temperature and nutrient profiles from several locations. Their data for silicate and phosphate agree well with those of Beauchamp (1939). They found a pronounced maximum in nitrate in the mid-thermocline between the base of the mixed layer and the top of the anoxic zone. This demonstrates that some nitrification/denitrification processes do occur in the lake. These mechanisms probably also play a very important role in the littoral zone.

Edmond in his systematic survey in 1973 sampled weekly at the two affluents of the lake, the Ruzizi and the Malagarasi. This study may be considered the most complete limnological study of the lake pelagic–sublittoral zone. Since he sampled the estuaries of the Ruzizi and the Malagarasi, some interesting conclusions can be drawn for the littoral and sublittoral zone. He registered lower nitrogen/phosphorus ratios in the southern compared to the northern part of the lake and concluded that the lake's productivity is nitrogen-limited and must depend to a substantial degree upon nitrogen fixation, both in the lake itself and by inputs from rainfall. (Rain conductivity was measured during 1992 by P. Coveliers, and never exceeded 6 μS/cm). The distribution of *Anabaena flos-aquae* may be indicative of the relative importance of nitrogen fixation in Lake Tanganyika at that time, as this species always had heterocysts present and could have been fixing atmospheric nitrogen. High rates of nitrogen fixation are required to balance the nitrogen budget of the lake, and the southern basin has a greater nitrogen deficit relative to the phosphorus because upwelling is confined to the southern end.)

Tietze *et al.* (1980) have demonstrated that in Lake Kivu very high CH_4 abundances occur in the anoxic zone dominated by bacterial metabolism of organic carbon. It appears that this is also the case in Lake Tanganyika, where the high carbon/nitrogen ratios in recently deposited sediment (Haberyan and Hecky, 1987) may reflect a similar preferential release of NH_4. Physical parameters and chemical parameters in terms of the major ions of the lake (Na^+, K^+, Ca^{++} and Mg^{++}) have been extensively measured during several missions. During his 1973 mission Craig collected data on CO_2, $^{13}C/^{12}C$, ^{226}Ra, ^{210}Pb, He and $^3He/^4He$ ratios, deuterium and ^{18}O, major ions, nutrients and Ba at various depths.

Well documented, and confirmed during several missions, is the thermal stratification of the lake, although some questions remain. A project of Finland's Department for International Development Co-operation with the Food and Agriculture Organization of the UN (FINNIDA/FAO) is focusing on the physical and chemical properties of the lake's pelagic zone. Among the findings of the FINNIDA/FAO project:

"The oxygen concentration, as well as the temperature at different depths are very important factors directly influencing all aquatic life forms. Nutrients like phosphate and nitrogen act as 'food' for flora and fauna which form the basis of the food chain. The other parameters mentioned in the method section (temperature, pH, conductivity, water visibility, and O_2 concentrations are measured from the boat, whereas the concentrations of phosphate, nitrogen compounds, Na^+, K^+, Ca^{++}, Cl^-, Si, SO_4^- and turbidity are determined in the laboratory upon return) give a more general picture of the lake's water quality and may give information on processes like upwelling and downwelling, mixing of water layers and other hydrodynamic events. It is true that all the parameters measured are interrelated, and that the processes which control these parameters [especially the nutrients] are important. Eventually this knowledge will enable us to provide predictions on the actual situation, abundance [and possible harvesting], production, etc. of the animals occupying positions higher up in the food chain, such as clupeids and perches (*sic*)." (Langenberg, 1994).

Apart from some observations by Edmond *et al.* (1993) and Cohen *et al.* (1993), no impact studies of excess sediment discharge into Lake Tanganyika's pelagic zone have been carried out.

The littoral and sublittoral zones

Gourdin *et al.* (1988) published the results of a study in which samples were analysed monthly during one year for temperature, pH, conductivity, suspended solids, Na^+, K^+, Ca^{++}, Mg^{++}, CO_3^-, HCO_3^-, Cl^- and SO_4^-. Samples were taken for Lake Tanganyika Bujumbura, Lake Tanganyika Gitaza, Lake Tanganyika Magara, Lake Tanganyika Resha, Lake Tanganyika Karonda, Lake Tanganyika Nyanza-Lac and the rivers Mutimbuzi, Ntahangwa, Kanyosha, Rusibasi, Rugata, Cugaro, Dama, Mulembwe, Buzimba, Nyengwe and Rwaba (a copy of the sampling locations is included).

Only very recently (within the last 5 years) has research begun on the impact of excess sediment discharge on the limnology of the littoral and sublittoral. During the 1992 expedition of A. S. Cohen, J.-J. Tiercelin and K. Martens, impact studies and physico-chemical analyses were carried out. Although this mission included only one season, and was therefore too short to make a comparison of the effects of sediment discharge on the food web during wet and dry seasons, some interesting results emerged. At high human-impacted deltas the bottom of the lake became anoxic with formation of H_2S and NO_2^-; low redox conditions prevailed.

Projects such as the UNESCO/DANIDA project and the CRRHA (Centre Régional de Recherche en Hydrobiologie Appliquée) project, both based in Bujumbura and running for longer periods (5 years), were able to present more comprehensive and global effects of excess sedimentation input on the lake littoral and sublittoral limnology.The UNESCO/DANIDA project (G. Ntakimazi, K. West, P. Coveliers and M. Zaliewski) studied the effects of three different river deltas (rivers with high human impact, low human impact jungle, and medium human impact wetland drainage basins). During dry season and wet season, sampling of the rivers as well as sampling in the delta was done in order to evaluate the impact of these rivers on the lake. Nutrients, major ions, dissolved organic carbon (DOC), and chlorophyll *a* were analysed, and fishing was done by gill nets. Rainfall was measured monthly.

The effect of excess sedimentation by riverine input appeared to have an effect on the water quality as well as on the biodiversity. Less biodiversity was found at the high human impact deltas, while for the physico-chemistry the nitrogen/phosphorus ratios changed dramatically. Transparency in the deltas of the high human impact rivers declined, high suspended solid loads were registered, and phosphate concentrations augmented. Chlorophyll *a* concentrations decreased, redox potentials decreased, and a shift in chemical speciation occurred. At the pristine (jungle ecotone) delta, pH decreased and chlorophyll *a* was augmented. These data confirm previous work by Shapiro (in press), in which he concludes that a lower pH gives green algae a competitive advantage over blue-green algae.

The CRRHA team sampled six rivers (Ruzizi, Kanyosha, Mugere and Ntahangwa for the Burundian side; Kalimabeja and Murongwe for the Zairian side) and their deltas monthly for nutrients, major ions, suspended solids and chlorophyll *a*. Fishing was done on a more irregular basis. The drainage basin of these rivers can be considered as high-to-medium human impact. P. Coveliers, K. West and A. Cohen sampled 12 rivers south of Bujumbura (Rwaba, Muguruke, Mukungu, Nyengwe, Mukunde, Kamango, Ruzibazi, Nyarhongo, Rutunga, Nyamusenyi, Kirasa and Karonga) monthly for pH, conductivity, oxygen, temperature, flux and turbidity. Occasionally redox potentials were measured. Three of these rivers (Rwaba, Kamango and Ruzibazi) were also used for the DANIDA/UNESCO research project.

Conclusion

The three studies undertaken by the CRRHA team, the UNESCO/DANIDA team and the expedition of Edmond *et al.* (1993, 1995) give some insight into the limnological aspects of the sediment discharge into the lake's littoral and sublittoral zones. Although they do not give a fully comprehensive insight into the problems, they nevertheless bring out some interesting aspects of the sedimentation pollution problems with regard to the lake and the lake basin's specific conditions. Effects on benthic organisms, sediment pore water (absorption/desorption processes) and microbiological fauna (nitrification/denitrification) were not analysed.

2.5 IMPACT OF SEDIMENT ON PRIMARY PRODUCTION IN LAKE TANGANYIKA

2.5.1 Introduction

Primary production has been found to be a good predictor of fish yield in a number of tropical and many temperate lakes, and Hecky *et al.* (1981) applied their 1975 estimate of annual primary production in Tanganyika, 290 g C/m^2/year, to equations used for other lakes. However, the results were again below current annual fish yields by wide margins. These authors concluded that the predictive equations appeared inappropriate for Lake Tanganyika, and an explanation of the higher-than-expected yields suggested (a) that the efficiency of carbon transfer to the fishery from primary production is anomalously high in relation to most lakes, and (b) that heterotrophic bacterial production, which appears equal to or greater than algal production, was somehow supplying the food chain supporting planktivorous fish. Therefore in order to evaluate the primary production in littoral and sublittoral zones, bacterial production should be involved.

2.5.2 Algae

Sublittoral areas

Most studies on primary production concern the pelagic area of Lake Tanganyika. The first direct measurements were carried out on one single station in the northern part in April 1971, and published by Melack (1980). The biomass production of the algae was rather low. Hecky *et al.* (1978), Hecky and Fee (1981) and Hecky and Kling (1981) have done further work on phytoplankton biomass and chlorophyll content in the pelagic zone. They found that the mean annual phytoplankton biomass for the whole lake was 140 mg/m^3, not at all a low algal biomass production as was first thought (they noticed that small chlorophytes were lost in the early research). This mean annual phytoplankton biomass was extrapolated from data of two intensely sampled stations (Bujumbura and Kigoma), and from two north–south transects of the whole lake. The station at Bujumbura was sampled 9 km offshore from February to November at 5-m intervals from the surface through to 25 m. The station near Kigoma was sampled 6 km offshore from March–October. Results from the station near Bujumbura showed three distinct seasonal phases:

- very low biomass (<100 mg/m^3), mid-February to end April
- somewhat higher biomass (100–200 mg/m^3), May to mid-September
- highest biomass (>200 mg/m^3, max. 920 mg/m^3), mid-September to mid-February.

The first period was dominated by chlorophytes (cyanophytes, diatoms), the second by chrysophytes (cryptophytes) and the third by diatoms and the cyanophyte *Anabaena* sp. Chrysophytes were dominant during the well stratified period. Similar succession was found in the station near Kigoma, but the biomass was on the average lower. An important source of food is not calculated in the above-mentioned biomass, because the protozoan biomass was on average nearly equal to the phytoplankton biomass. A green alga, living symbiotically in the ciliate *Strombidium* cf. *viride*, is very probably responsible for this phenomenon. It was estimated that 75% of the total protoplasm biomass is due to this green alga, and only 25% to *Strombidium* itself. The two north–south transects of the whole lake taken in 1975 demonstrate a higher algal biomass in the northern part of the lake during October–November (Hecky *et al.*, 1978). The algal biomass readings are highest near the inlet of the Malagarazi river during April–May (eutrophication?); the other sampling stations, in the northern and southern basins, have a similar biomass production.

Littoral zones in the northern part of Lake Tanganyika, along the Burundian coast

Diatoms are often the most important algal group in the littoral (Cocquyt *et al.,* 1991). Suspension of surface sediment organisms can be induced in the phytoplankton of the littoral by the often strong wind action along the coastline of the lake and by local circulation of the water column due to input of river water. Therefore the phytoplanktonic diatoms can be considered as tychoplanktonic. The greater number of algae in the littoral near Bujumbura and the abundance of Euglenophytes in this station, point to a level of pollution (eutrophication). The data on the seasonal and spatial phytoplankton distribution on the littoral along the Burundian coast (Cocquyt *et al.,* 1991) are expressed in cells/ml. It is possible to convert these data sets to biomass expressed in mg/m^3, although it would require a huge amount of work. Conversion is also possible of the data on phytoplankton composition in the Bay of Bujumbura (Caljon, 1992). K. Wijonda recently carried out work on phytoplankton in the littoral zone of the lake, in which he compared the phytoplankton populations at three different delta areas (Ntahangwa, Mugere and Kanyosha-Mugere); unfortunately these are all rivers with a human-impacted drainage basin.

2.5.3 Bacteria

Van Meel (1954) suggested that bacteria could play an important role in the trophic dynamics of Lake Tanganyika. He invoked bacterial production, without any observational data, to resolve the paradox of the concurrence of high transparency, sparse phytoplankton and abundant zooplankton. Hecky and Kling (1981) showed that bacteria were reasonably abundant in the lake. Important from a trophic point of view is the utilization of dissolved organic carbon by bacteria. This converts fixed energy from the dissolved form to the particulate form which can then be ingested by larger organisms. For microbiological growth to be an important trophic pathway rather than a dead end dissimulating fixed carbon, there must be organisms harvesting bacteria. The protozoans are a prominent feature in the pelagic of Lake Tanganyika. Their biomass is often comparable to that of the phytoplankton. If the above-mentioned symbiosis between a green alga and *Strombidium* cf. *viride* is not at least the major factor for the equality of biomass between protozoans and phytoplankton, then bacteria are the only remaining explanation for this phenomenon.

The 'microbial loop' by which dissolved organic carbon (DOC) enters the food chain has been emphasized in the marine pelagic and it very probably plays an important role in Lake Tanganyika (Hecky, 1991). Hecky *et al.* (1978) studied bacterial densities, but only in the pelagic of Lake Tanganyika. They found that bacterial densities across the interface were similar to those occurring in the epilimnion. On the west coast, bacteria form white, yellow or reddish-brown underwater filamentous mats in the vicinity of Pemba (Tanghydro Group, 1992). Hydrothermal activity, sulphide deposits, methane and hydrocarbon gases are present at this locality.

2.5.4 Importance of primary production in the littoral zone for the food chain (with reference to fish)

The primary production in the littoral zone is very important. Of the 287 known fish species from Lake Tanganyika, 207 species are known from the littoral and sublittoral (Coulter, 1991b). The economically important fish species *Stolothrissa* and *Limnothrissa*, each have an inshore phase in their life cycle, though both are pelagic in Tanganyika. Other species (e.g. *Lates angustifrons*, *Boulengerochromis microlepis*, *Dinotopterus cunningtoni*) are represented in the littoral by day. Certain polyspecific genera are primarily littoral and live on detritus or unicellular algae, micro-invertebrates and/or diatoms of the benthos, while in the bathypelagic they live on plankton. Diatoms constitute a major part of the diet of the so-called 'pickers'. A certain proportion of the littoral fish species are also direct consumers of the phytoplanktonic (tychoplanktonic) and benthic algae, though the precise percentage is not known at present.

2.5.5 Conclusion

Studies on primary production and biomass have been undertaken only in the pelagic of Lake Tanganyika. Investigations on algal and bacterial productivity in the littoral of the lake have not so far been undertaken. The stations near Bujumbura and Kigoma could be useful for comparing primary production and biomass with data from the littoral zone of the lake (the stations Bujumbura and Kigoma are situated in, or not far from, the sublittoral).

Primary production is much higher in the northern than in the southern basin. The most important productivity appears to take place at the beginning of the wet season. The importance of bacteria in the food chain is not yet solved. If the cyanophytes are considered to be bacteria (domain Eubacteria), then bacteria are very important for primary production in the lake.

The impact of excess sediment discharge on primary productivity, nitrogen fixation (e.g. *Anabaena* spp.), nitrification, denitrification, phosphorus uptake and release, organic carbon processing, etc. are unknown. The effects of excess sediment discharge on the distribution of bacterial and algal species in littoral and sublittoral zones are also unknown.

2.6 IMPACT OF SEDIMENTS ON SECONDARY PRODUCTION (AND TRANSFER OF CARBON BETWEEN PELAGIC AND BENTHIC ECOSYSTEMS)

2.6.1 Pelagic ecosystem

The pelagic waters of Lake Tanganyika are considered to be highly efficient in transfer of energy and carbon between trophic levels (Hecky, 1981). This may involve both an efficient phytoplankton–grazer–fish community, and utilization of microbial production, perhaps associated with the relatively shallow oxygen boundary which enables a transfer of nutrients from the deeper waters via a bacterial and protozoan food chain. Turbidity-induced changes in feeding capability, growth rates or production of zooplankton (including ciliates), or predation efficiency of planktivorous fish, will alter carbon fluxes through the pelagic food web. Alterations in community trophic dynamics could be expected to lead to shifts in the relative abundance of species, which may or may not be considered serious depending on whether or not 'keystone' species are affected. Whilst the open-water pelagic zone is least likely to be directly affected by catchment-derived turbidity, changes on a community basis in the inshore zones of the lake could be of major significance to the offshore ecosystem. For example, both of the clupeid fish in the lake have an inshore phase to their life history (Coulter, 1991b) with clear implications for offshore populations. The pelagic fish species *Lates mariae*, *L. microlepis* and *L. angustifrons* spend time as juveniles inshore, particularly among macrophytes (Coulter, 1991b). Changes of both food supply and habitat, through, e.g. association of macrophytes and sediment structure (Ali *et al.*, 1995), could have consequences for juvenile survival and thus for recruitment to the offshore adult populations.

2.6.2 Benthic ecosystem

Benthic productivity is dependent on the supply of nutrient and organic matter from overlying water. This may be passive, through sedimentation or active, involving biotic transfers. Whilst the benthos can be considered as an ultimate sink for carbon, there is also a continual process of transfer back and forth between the sediment and water. This may involve wind- or current-driven resuspension, or movement of organisms into and out of the sediment and sediment/water interface. Low specific gravity particles generally both have a higher nutritive value than high specific gravity particles and remain in suspension for longer periods (Bowen, 1984). This is of relevance to potential bioavailability.

Changes in the productivity within the water column translate to the benthos (Johnson *et al.*, 1989). This may be direct through epiphytic or epilithic fixing of carbon by algae and bacteria, or indirect, either through a detrital rain, subsequently mineralized or through feeding by benthic organisms. Increased turbidity in the water column would reduce light penetration and therefore the areas of benthos on which photosynthesis can occur. Benthic primary production is likely to be a major contributor to total primary production in the relatively shallower regions of the lake as it is in the southern end of Lake Malawi (Bootsma and Hecky, in press). However, it is also possible that in shallow water, and with sufficient photosynthetically active radiation reaching the benthos, flocculation of sediment and phytoplankton increases settlement and benefits epiphytes by improving light climate and nutrient availability (Burkholder and Cuker, 1991). Phosphorus bound to sediment may enhance associated benthic algal growth, and anaerobic microclimates can increase phosphorus availability (Lock *et al.*, 1984; Riber and Wetzel, 1987).

Changes in the abundance or nature of the benthic primary producers are likely to have consequences for populations of secondary producers. This may be more likely in a water body such as Lake Tanganyika with a variety of endemic and probably highly specialized species. The inshore littoral and shallow benthic zones have a rich and diverse fauna (Coulter, 1991a). This includes a diverse ostracode and mollusc fauna whose endemic speciation is in many ways as impressive as that of the cichlid species flocks of the African Great Lakes. The molluscs are thought to have co-evolved along with the endemic potamonautid crabs (West and Cohen, 1994).

Organic allocthonous inputs to the benthos are important for a variety of detrital feeders in Lake Tanganyika: oligochaetes, benthic copepods and ostracodes, atyid shrimps and insects. Collectively these comprise a multitude of species; many of which are endemic and most of which are confined to the nearshore littoral. Energy conversion efficiencies of this community in the lake are unknown; indeed there have been relatively few studies which have investigated energy and matter transfer into the benthos (Wetzel, 1983; Jónasson *et al.*, 1990; Martinet *et al.*, 1993). Martinet *et al.* (1993) concluded that 4.4–6.4% of allochthonous organic carbon input was converted into tubificid biomass in the backwater of the River Rhone. However, of perhaps greater significance to the ecological threat from increased sediment load in Lake Tanganyika, they also noted a shift in benthic community structure favouring more *r*-selected species with increased carbon sedimentation rate and/or benthic instability.

Changes in the quantity or frequency of pulses of organic detrital material to the benthos, whether inshore or in the deeper profundal, will be reflected in production of benthic invertebrates and in the fish community which feed upon them. The ecological response to increased inputs of suspended solids, both in the inshore pelagic and the benthos, will depend on the nature of the sediment and precise spatial and temporal distribution patterns. It will also depend how much organic material and nutrients forms part of the sediment load. That will be related to the organic and nutrient status of catchment soils, perhaps situated many kilometres away from the lake basin. The increasing realization of the general importance of detritus on the trophic ecology of aquatic systems (Wetzel, 1995) suggests that, regardless of other effects of turbidity, changes will or have already been induced in the carbon transfer systems of Lake Tanganyika. Consequently the resistance of food webs stabilized through co-evolution in the lake (Matsuda and Namba, 1991) may have already been tested.

2.7 IMPACT OF SEDIMENT ON THE ALGAL SPECIES OF LAKE TANGANYIKA

2.7.1 Algal populations and species distribution

Introduction

Hecky *et al.* (1978) and Hecky and Kling (1981) investigated the phytoplankton composition of the euphotic zone of Lake Tanganyika. Data from two intensely sampled stations (Bujumbura and Kigoma) and two north–south transects of the whole lake were studied. Chlorophyta are the most important group over almost the entire sampling period (February–November) in the station near Bujumbura. Seasonal succession was observed in Cryptophyta, Chrysophyta and diatoms. Cyanophytes predominate from October until March. The phytoplankton composition near Kigoma differs from the station near Bujumbura: cyanophytes are more numerous, and the seasonal succession looks different between the Chrysophyta and Cryptophyta. Data from the two north–south transects show that the Chlorophyta are the most abundant group in the southern basin during the period October–November (in most sampling stations). Cyanophytes, on the other hand, seem to be more abundant in the southern basin during April–May; Chrysophyta are predominant in the whole lake during this period. Diatoms, on the other hand, are very significant only in the northernmost part of the lake. The proportional composition of different distribution types of diatoms (only the names of the most important genera are mentioned) is given by Cocquyt and Vyverman (1994). Distribution patterns for this group are the best documented. 69.8% of the diatoms reported from Lake Tanganyika have a cosmopolitan distribution: 4.8% are pantropical; 10.6% are African and 14.8% are tropical African taxa. 8% of the total diatom flora (30 species, of which 13 belong to the genus *Surirella*) have a distribution restricted to Lake Tanganyika. Endemism of algal taxa is not at all common, compared with zoological organisms. Studies on endemism and cosmopolitanism in the diatom flora of East African Great Lakes were carried out by Ross (1981), who concluded that the distribution patterns show that many species of diatoms have very great ecological tolerance, that they are capable of long-distance dispersal and that speciation in general is low. He could not give any explanation for the much greater rate of speciation in the Surirellaceae (*Surirella* and *Cymatopleura*).

Littoral zone in the northern basin

The spatial distribution of phytoplankters has been studied in the Bay of Bujumbura (Caljon, 1992), and along the north-eastern coast of Lake Tanganyika (Cocquyt *et al.*, 1991). As in the pelagic (Hecky *et al.*, 1978), the algal populations in the littoral undergo seasonal variations. Phytoplankton composition in the littoral mainly differs from the pelagic in the greater diversity of diatoms, while Dinophyta and Chrysophyta seem to be better represented in the pelagic. The greater importance of diatoms in the littoral is probably due to local influences, e.g. suspension of diatoms of surface sediments induced by the often strong wind action along the coast (tychoplankton), and local circulation of the water column due to the input of river water (Cocquyt *et al.*, 1991). Euglenophyta are more important near Bujumbura, caused by higher nutrient values (eutrophication) (Caljon, 1991; Cocquyt *et al.*, 1991).

The study undertaken in the Bay of Bujumbura (Caljon, 1992) reports the species names of all groups. The other study (Cocquyt *et al.*, 1991) is restricted to the species names of the most important diatoms.

Conclusion

To date, limited studies have been made of algal populations and species distribution in the littoral of the lake, and solely along the Burundian coast line. The sampling stations situated near the sublittoral can be useful for comparing algal composition in the littoral zone of the lake. In the pelagic of the lake, algal composition differs between sampling stations and during the year.

The presence of endemic algal species in Lake Tanganyika makes this lake particularly interesting. Endemism is rather rare among phytoplankton. At present, nothing is known about the distribution of these endemics in Lake Tanganyika.

2.7.2 The effect of increased sediment loads on algal species diversity

Introduction

The first reports on algae of the East African Great Lakes date back to the end of the last century (e.g. Dickie, 1880; Schmidle, 1898). Contributions restricted to Lake Tanganyika were published by West (1907). A survey of all the taxonomic work up to the middle of the 20th century is given by Van Meel (1954). Renewed interest in taxonomy and biodiversity during the last decades continued the earlier systematic investigations, mostly focused on diatoms (e.g. Caljon, 1987; Kociolek and Stoermer, 1991, 1993; Cocquyt, 1991; Caljon and Cocquyt, 1992). An updated checklist of all algal taxa hitherto reported from Lake Nyasa/Malawi, Lake Tanganyika and Lake Victoria has been published (Cocquyt *et al.*, 1993). Nomenclatural and taxonomic revisions of the algal flora during the last years have resulted in important changes of taxa. Therefore special attention was paid to possible synonymy in the checklist. However, the earlier reported species were not checked with type material. Misidentification of taxa in earlier studies, revealed by this renewed taxonomy (e.g. *Stephanodiscus astraea* complex) still remains a problem. Cocquyt and Vyverman (1994) mentioned the number of infrageneric taxa and genera per major taxonomic group recorded from Lake Tanganyika, together with data from Lake Malawi and Lake Victoria. Diatoms are the most important group with 474 infrageneric taxa, followed by the Chlorophyta (224 taxa, including five Charophyta), Cyanophyta (111 taxa), Euglenophyta (59 taxa), Chrysophyta (21 taxa), Dinophyta (19 taxa), Cryptophyta (14 taxa), Xanthophyta (four taxa) and the Prymnesiophyta (one taxon). The total number of algal taxa reported to date from the lake is 927, belonging to 184 genera. Van Meel (1954) reported only 241 taxa and Coulter (1991a) 578 taxa. The increasing numbers of taxa are due to the recent studies in the northern part of the lake. Most of these recent studies concern diatoms: Van Meel (1954) reported 103 diatom taxa; Coulter (1991a) 220 taxa (see Cocquyt and Vyverman, 1994). The same trend is seen for the Chlorophyta: Van Meel (1954) reported 79 taxa; Coulter (1991a), 149 taxa; and Cocquyt *et al.* (1993) revealed 224 taxa.

Algal species diversity in the pelagic zone

Most of the earlier work on diatoms was done on the pelagic or in the sublittoral. No direct figures can be given on the algal species diversity in the pelagic, but one thing is clear: the littoral is more diverse.

Northern basin

Kufferath (1956) was the first to examine sediment cores from the northern part of Lake Tanganyika. Studies on two sediment cores taken in the Bay of Burton were carried out (Mondeguer *et al.*, 1986). Only one algal taxon, *Nitzschia spiculum*, is mentioned by its species name.

Diatoms in 49 surface sediment samples, taken on four east–west transects in the northernmost part of the lake, were studied (Caljon and Cocquyt, 1992). All analysed material was treated with acetic acid and hydrogen peroxide. No living material was investigated. 36 samples were taken in the pelagic, nine in the sublittoral and two in the littoral. 227 diatom taxa were distinguished. Based on the literature, the authors concluded that 4% of the organisms were planktonic, 29.5% preferred planktonic and benthic habitats, and 54.6% preferred aerophytic or benthic habitats. The other 11.9% could not be determined.

Symoens (1955a, 1955b, 1956, 1959) described an *Anabaena* bloom at the end of the dry season.

Southern basin

Studies on a sediment core taken near the southern end of the lake, consisting of 1060 cm of sediment, give an idea of the diatom composition during the last 14 000 years (Haberyan and Hecky, 1987).

Algal species diversity in the littoral and sublittoral zones

Northern basin

North and east coasts. Cocquyt *et al.* (1991) give a species list of diatoms taken at five stations along the Burundian coast. In addition, they mention the presence of 89 other taxa belonging to the Cyanophyta (35), Chlorophyta (30), Euglenophyta (11), Cryptophyta (7), Chrysophyta (3), Xanthophyta (2) and Dinophyta (1). Algal taxa from 12 samples taken in the Bay of Bujumbura are given by Caljon (1992). Two littoral and nine sublittoral samples of surface sediments, taken on an east–west transect in the northernmost part of the lake, were investigated together with the above-mentioned pelagic samples (Caljon and Cocquyt, 1992). Mpawenayo (1985, 1986) also mentions some diatoms in the northernmost part of the Lake. Unpublished honours theses of final-year students at the University of Burundi, conducted by A. Caljon, deal with phytoplanktonic and benthic research in the littoral along the Burundian coast. At present, a collaborator of the Belgian CRRHA project is working on planktonic samples taken in the vicinity of Bujumbura. Cocquyt (1991) described a new diatom taxon from thrombolytic reefs of the sublittoral in the south of Burundi. 104 diatom species were found in two samples from this reef.

West coast. There have been few systematic studies on the algal flora along the Zairian coastline. Only one publication in recent literature deals with data on planktonic and benthic algal organisms (Muzino, 1987). Muzino reports about 14 diatomic and 17 non-diatomic taxa from sample stations situated near Uvira and Luhanga. Moreover, most taxa are not determined at species level.

Southern basin

No data other than those mentioned in general works are available from the littoral of the southern basin.

Conclusion

During the 1980s to 1990s, studies on species diversity have been undertaken more intensively. The Bay of Bujumbura and the Burundian coastline have been the focus of these investigations. The increasing number of reported taxa from Lake Tanganyika from investigations of such a small area of the whole lake points to the need for further research. Not much attention has been paid to the great variety of habitats, e.g. samples on rocky coastlines have yet to be closely examined.

2.8 IMPACTS OF SEDIMENTS ON THE MACROPHYTES OF LAKE TANGANYIKA

2.8.1 Introduction

Only 18 macrophyte taxa have been reported from Lake Tanganyika (Coulter, 1991a). The most common genera are *Ceratophyllum*, *Potamogeton*, *Vallisneria*, *Myriophyllum*, *Najas* and *Chara*. Of the Charales reported from the lake (*Chara* and *Najas*), *Chara setosa* f. *setosa* and f. *tanganyikae*, *C. vulgaris* f. *gymnophylla*, *C. zeylanioca* and *Nitella mucronata* (Cocquyt *et al.*, 1993), only *Chara setosa* f. *tanganyikae* is endemic to the lake, the other taxa being more commonly distributed throughout tropical Africa.

Due to the steep inclination of the generally rocky lake-bottom, the macrophytes are not abundant in the littoral and their distribution is discontinuous. Exception has to be made for some areas, for example, some sheltered shallow regions (e.g. near the islands of Kipili, Tanzania and at Mpulungu, Zambia); areas where sandbanks are parallel to the shore or large estuaries where sediment cones occur (e.g. Malagarazi estuary, Tanzania); shallow protected bays (e.g. Burton Bay, Republic of Congo) (Coulter, 1991a).

Najas pectinata and *Ceratophyllum demersum* are characteristic of the larger, calm bays in the northern parts of Lake Tanganyika (Lewalle, 1972). The same author mentions the importance of submerged aquatic vegetation in the vicinity of the Ruzizi delta. Coulter (1991a) reports 63 other macrophyte taxa in the lake. The great majority of macrophytes has been reported from lagoons, swamps and rivers. In the past, several authors studied the macrophyte vegetation along the coastline (e.g. Germain, 1952; Lewalle, 1972; Reekmans, 1985). Sandy beaches along the Burundian coast have a vegetation primarily composed of *Phragmites mauritianus*, *Cyperus papyrus*, *Typha domingensis* and *Vossia cuspidata*. As in the case of phytoplankton and phyhobenthos in the littoral zone, sandy beaches of the northern part of the lake have been studied more intensively than those of the southern part. However, during the last decade no further work has been done on the vegetation of the sandy beaches.

2.8.2 Conclusion

Studies on the impact of sedimentation on the macrophytes of Lake Tanganyika are limited. The importance of macrophytes and marginal vegetation as refuges for larval stages of fish is not clearly known, but this may be an important factor when considering the biodiversity of fish.

2.9 EFFECT OF INCREASED SEDIMENT LOADS ON ZOOPLANKTON

2.9.1 Introduction

Anthropogenic accelerated soil erosion is a major global environmental problem. In semi-arid and humid tropical areas erosion rates may be particularly severe. Typical rates of erosion are estimated as 10–50 t soil loss/ha/year in sub-Sahel West Africa (Lal, 1993) and up to 70 t/ha/year in the highlands of Ethiopia (Hurni, 1993). In particularly vulnerable areas such as deforested slopes or areas ploughed along the gradient of a slope, rates can be considerably higher (Fournier, 1967; Roose, 1977a,b; Lal, 1986; El-Swaify, 1993). Whilst the erosion potential is dependent on geomorphic characteristics such as soil type, elevation and rainfall pattern, the realization of that potential is clearly related to patterns of human land use such as overgrazing, forest clearance and agricultural practice. As such, whilst the global scenario of land degradation makes for sober reading (Pimental, 1993), most problems can be reversed by more rational land management made possible by educational and conservation measures. In addition, the link between population increase and land erosion in parts of the world needs to be understood (e.g. Craswell, 1993).

Soil erosion is usually caused by water run-off, which often involves a mechanism of positive feedback (Graetz, 1991). Eroded sediment inevitably transcends to rivers which flow to lakes, reservoirs or the ocean. Catchment sediment transport is difficult to quantify because rates of movement depend on particle size and the hydrology and geomorphology of the river basin. Fine particles, organic matter, silts and clays are more susceptible to erosion and subsequent transport than coarse particles (Evans and Skinner, 1987). Whilst transport of organic matter can be rapid it is not constant and eventual deposition may occur long distances from the source (Cushing *et al.*, 1993). The prevalence of organic matter is greater in fine material than in coarse, and the susceptibility to erosion and transport of these fine particles is significant both for nutrient loss from lands and to gain to freshwaters. Phosphorus and some pesticides bind preferentially to fine particles (Arden-Clarke and Evans, 1993) and organic matter is often mineralized during water-borne transport. Stocking (1984) estimated an annual erosive loss in Zimbabwe of 236 000 t phosphorus and 1635 000 t nitrogen; equivalent to $1.5 billion in fertilizer. Cullen and O'Loughlin (1982) estimated average losses from eroded Australian soils of 0.4 kg phosphorus/ha/year.

2.9.2 Lake Tanganyika

Sediment erosion from land and transport to the lake basin has been identified as a form of pollution affecting the biodiversity of Lake Tanganyika (Cohen *et al.*, 1993). The degree of sediment load to the lake and the extent of lake area affected need further elucidation. It is, however, important in such estimations to recognize and attempt to quantify lag times between land loss and lake gain. For river sediment transfer this can be 1–10^2 years (Newson, 1992). Although actual delivery rates of sediment to the lake are unknown, it is not unreasonable to consider that these will have increased in recent decades in response to increased pressure on land use in the catchment. The effect of increases in suspended solids and subsequent settlement on the planktonic and benthic communities, and on carbon transfer within and between the pelagic and benthos, is dependent on the nature of the sediment, its temporal and spatial pattern and the relative susceptibility of impacted organisms. Sediment load into Lake Tanganyika will be related to seasonal rainfall pattern in the catchment and, at least to some extent, the prevailing thermal structure of the lake. During periods of stratification, incoming water tends to enter directly into the hypolimnion and may have a limited effect on biotic processes in the overlying epilimnetic water. The susceptibility of biota to sediment load is variable and it is therefore important to focus any study on the specifics of the Tanganyika communities, rather than a perhaps unrealistic reliance on extrapolations from studies done elsewhere.

2.9.3 The zooplankton communities of Lake Tanganyika

Offshore pelagic community

The invertebrate communities of the lake are reviewed by Coulter (1991a). For the zooplankton there is a clear distinction between organisms of the pelagic and those of the littoral zones of the lake. The pelagic invertebrate community comprise protozoans, cnidaria, copepods and decapod shrimps mainly of the family Atyidae. Cladocerans and rotifers, typical of many lacustrine environments, are rare or absent. This assemblage of organisms is unique and comprises a number of endemic species, particularly among the shrimps; 14 out of 15 recorded species are endemic. Hecky *et al.* (1978) identified 16 species of protozoans, dominated by the genera *Strombidium*, *Didinium* and *Coleps*, in the pelagic. The biomass of these can exceed that of phytoplankton during the wet season period of maximum stratification (Hecky and Kling, 1981). The commonest, *Strombidium viridae*, contain algal cells suggestive of a symbiotic relationship. The species of cnidarian in the lake, *Limnocnida tanganyicae*, is widespread but noted for the sporadic occurrence of large swarms. Its role in the trophic structure of the lake is unknown. The major converter of primary to secondary production in the lake is probably the calanoid copepod *Tropodiaptomus simplex* which was described by Dumont (1986) as comprising over 95% of the grazing zooplankton. Its adaptations to the generally food-limited waters of the lake, with chlorophyll *a* concentrations <3.0 µg/l and bacterial concentrations probably below the threshold for diaptomid maintenance by filtering (Lehman, 1988), are probably similar to those of *Tropodiaptomus cunningtoni* in Lake Malawi (Hart *et al.*, 1995). However, recent work has indicated that the relative abundance of calanoid to cyclopoid populations is greater in the south compared with the north of the lake (H. Kurki, FAO Lake Tanganyika Research, Kigoma, personal communication), which may explain generally lower phytoplankton abundance found in the south (Hecky *et al.*, 1978). Cyclopoid populations in the pelagic are dominated by the predatory *Mesocyclops aequatorialis aequatorialis* and the herbivores *Microcyclops cunningtoni* and *Tropocyclops tenellus*. Quantification of relative abundance of cyclopoid species appears not to have been included in previous or ongoing studies. Atyid shrimps are found in the pelagic but are probably more abundant inshore and associated with the benthos, although they have also been found in the stomach contents of pelagic fish species (Marlier, 1957).

Inshore littoral community

The inshore zooplankton is more diverse, and more endemic, than that offshore. This is most likely because of increased habitat structure inshore compared with the offshore zones. Many inshore ‘zooplankton’ will be associated with littoral growths of macrophytes or with the sediment–water interface. Species found offshore are generally also found inshore, although *Tropodiaptomus simplex* appears to be restricted to an offshore habitat and some protozoans appear to be more prevalent offshore. Cladocerans and rotifers are also found among the inshore zooplankton, but these may generally stray from associated wetlands and inflowing rivers. The diversity of inshore cyclopoids is particularly noteworthy, comprising 39 species. Many of these, such as the order Harpacticoida, will be strongly associated with the benthos. Others such as the Ergasilidae and Lernaeidae are fish parasites.

2.9.4 Effect of increased sediment load on zooplankton communities

Direct effects

Increased concentrations of suspended solids can affect zooplankton communities directly or indirectly. The main direct effect is that suspended solids interfere with zooplankton feeding or assimilation rates (Aruda *et al.*, 1983; McCabe and O'Brien, 1983; Koenings *et al.*, 1990; Kirk, 1991) and cause starvation or reductions in growth rates and fecundity (Kirk and Gilbert, 1990; Hart, 1988; Kirk, 1992). Increased concentrations of fine particles have a greater effect on generalized filter feeders or those which feed on small particles (microfilters) compared with animals feeding on larger particle size (macrofilters) or those able to select and grasp individual particles. The generalized model of cladocerans as microfilters and copepods as macrofilters (Friedman, 1980; Horn, 1985a,b) can explain the greater susceptibilities of cladocerans to increased sediment loads compared with other zooplankton groups (Hart, 1986, 1988; Kirk and Gilbert, 1990; Koenings *et al.*, 1990; Kirk, 1992). However, whilst copepods may be relatively less susceptible to suspended solids compared with an organism such as *Daphnia*, they are not immune from a direct reduction of feeding rates caused by increased turbidity (Hart, 1988; Jack *et al.*, 1993). The prediction of relative impact on different zooplankton species may be possible through a knowledge of grazing rate spectra (Gliwicz, 1977), the prediction of overall grazing rates (McCauley and Downing, 1985) and the size frequency and abundance of suspended solids. Whilst the crustacean zooplankton groups which are most susceptible to increased suspended solid concentrations are not prevalent in Lake Tanganyika, it would be expected that the selective feeding (macrofiltering) calanoid would make *T. simplex* more vulnerable to increased sediment load compared with the generally raptorial feeding and often predaceous copepods. Koenings *et al.* (1990) speculated that the calanoid *Diaptomus* was more susceptible than the cyclopoid *Cyclops columbianus* to high levels of inorganic turbidity. Whilst most reports focus on the detrimental effects of increased turbidity on zooplankton feeding, there is evidence from Cahora Bassa, an impoundment on the Zambezi River, that zooplankton populations can themselves reduce concentrations of suspended solids through feeding (Gliwicz, 1986). Small inorganic particles, if coated or aggregated with sufficient organic matter, can provide a suitable food for grazing zooplankters (Marzolf, 1980; Lind and Dávalos-Lind, 1991).

Whilst it has been shown that suspended solids can exert a direct (and usually negative) effect on crustacean zooplankton feeding and that, overall, cyclopoid copepods appear to be the least susceptible, it is impossible to estimate details of relative effects on the cyclopoid community of Lake Tanganyika. It may be a reasonable assumption that predaceous cyclopoids may be less affected than herbivorous ones, but as all cyclopoids are herbivorous for at least part of their life cycle it is impossible to provide any estimate of relative species impact. That would depend on the details of and differences among the cyclopoid feeding mechanisms and the exact nature of the sediment load (Cuker and Hudson, 1992). Various life stages of individual species will also respond differentially to increased concentrations of suspended solids. The prevalence of increased sediment loads in the inshore areas arising from river deposition and wind resuspension in the shallower depths would presumably make zooplankton confined to the inshore and littoral regions of the lake more susceptible than those which inhabit the offshore pelagic zone.

The ecology of the atyid shrimps in Lake Tanganyika is poorly understood. A detailed description of the feeding mechanism has been made for the genus *Caridina* by Fryer (1960) who categorizes it as a "microphagous chelate raptorial feeder" because it seizes food in a wider sense than 'raptorial' and feeds on small particles. The habitat of the atyid shrimps in Lake Tanganyika is mainly bottom-dwelling, although some species may be important in the trophic ecology of the pelagic (E. Bosma, FAO Lake Tanganyika Research, Mpulungu, personal communication). The habitat of the shrimps is therefore often strongly associated with sediment and they would thus be expected to be able to cope with high concentrations of suspended solids. The feeding ecology of *Limnocnida tanganyicae* is also uncertain. Like other cnidaria, it would be expected to feed on small zooplankton (Dodson and Cooper, 1983), but how it would be affected by increased concentrations of suspended solids is uncertain. A variable response to increased concentrations of suspended solids has been demonstrated for ciliate protozoans (Jack *et al.*, 1993).

Indirect effects

The potential effect of increased sediment load on zooplankton populations is not only related to possible interference with feeding mechanisms resulting in lower growth rates, fecundity and survivorship. Indirect effects may be of equal or greater importance. Turbidity alters the underwater light climate of a lake, increases light attenuation and can restrict phytoplankton production through light limitation (Knowlton and Jones, 1995; Lind *et al.*, 1992). This can cause shifts in algal community structure (Cuker, 1987) as well as reducing overall production. However, it is also possible that increased sediment load can promote algal production if that load is associated with significant nutrient inputs. However, during such episodes chemical- and biotic-mediated flocculation of fine sediment can also occur, leading to rapid settlement of particles. Changes in phytoplankton abundance or community have obvious potential effects on herbivorous zooplankton. If phytoplankton production is enhanced owing to increases in sediment-associated nutrients, then the production of zooplankton species (particularly those which can cope with high suspended solid concentrations) would be expected to increase also. On the other hand, increased sediment load can, through absorption, reduce available phosphorus in the water column to levels below those predicted from catchment nutrient loads (Jones and Bachmann, 1976, 1978).

Increased turbidity may also restrict visibility for zooplanktivorous fish, affect predator-size selectivity (Vinyard and O' Brien, 1976; Gardiner, 1981; McCabe and O'Brien, 1983) and lead to changes in zooplankton survivorship, competitive ability and, ultimately, zooplankton community structure. In Lake Tanganyika, the cladocerans which would be likely to respond to a turbidity-mediated release of predation pressure are absent. Size-selective theory (Brooks and Dodson, 1965) would predict that the largest zooplankton would respond best to a reduction in predation pressure from visual feeders, which in Lake Tanganyika would be *Tropodiaptomus* and *Mesocyclops*. However, these organisms are already of a very small size compared to most of those animals which have featured in the fish–zooplankton predator–prey investigations, and the theory may or may not hold for such size ranges. There are also the complications of variable copepod escape ability (Drenner *et al.*, 1978) and predation refugia already offered by littoral sediment and macrophytes (Timms and Moss, 1984).

2.10 EFFECT OF INCREASED SEDIMENT LOADS ON BENTHIC INVERTEBRATES

2.10.1 Extant biodiversity of Lake Tanganyika

Tabulations by Coulter (1991a, 1994) have shown that the extant faunal diversity in Lake Tanganyika totals approximately 1300 described species, about half of which are endemic. Percentage endemicity can vary substantially between groups. The highest diversities are found in the littoral. Table 2.5 gives extant diversities for benthic invertebrate groups (from Coulter, 1994). This means that approximately one third of the total extant diversity in Lake Tanganyika is formed by benthic invertebrates; furthermore, most of these are living in the littoral and are thus most susceptible to environmental impacts by excess sedimentation. Recent advances in the taxonomy and phylogeny of benthic invertebrate groups (Martens, 1984; Wouters and Martens, 1992, 1994; Martin and Brinkhurst, 1994; Martin and Giani, 1995a,b) indicate that this fraction is undoubtedly much larger.

2.10.2 Impact of excess sedimentation

The possible effects, direct or indirect, of excess sediment loading in the water column (and on the bottom) on aquatic invertebrates have been listed by Cohen *et al.* (1993b,c); these papers also present an exhaustive list of references on the subject. The following possible effects can be distinguished:

- reduction of light penetration, hence of photosynthetic rate and possibly other environmental variables (such as temperature);
- blanketing of benthic algae by settling sediments;

Table 2.5 Extant diversities for benthic invertebrate groups (from Coulter, 1994)

Order	Genera	Species	Endemic species
Nematoda	12	20	7
Oligochaeta	7	8	5
Hirudinea	8	20	12
Gastropoda	36	60	37
Bivalvia	10	15	9
Copepoda	23	69	33
Ostracoda	16	84	60
Decapoda	6	25	22
Insecta	3	107	107
Total	**121**	**408**	**294 (72%)**

- reduction of nutritional value of detritus, possibly affecting nutrient dynamics of entire water bodies;

- reduction of nutrients from water column by sedimentation;

- possible changes in food supply caused by the latter two effects as the nutrient/plankton relationships change;

- physical damage, i.e. to feeding apparatus of filter-feeding organisms;

- reduction of habitat complexity (filling in of crevices, reduction of total rock surface).

Of these possible effects, the reduction of habitat complexity and the physical damage to feeding apparatus appear to be the most important for benthic invertebrates.

2.10.3 Current knowledge for Lake Tanganyika

A literature review of the impact of excess sedimentation on aquatic organisms is presented by Cohen *et al.* (1993). Earliest mention in the literature are by Bizimana and Duchafour (1991) and Nsabimana (1991). The former authors cite a soil loss of 27 t/ha/year on slopes of 28% up to 100 t/ha/year on slopes of 49% under traditional manioc cultivation. Results of various periods of field work (1985–89) were used by Cohen (1991b, 1992) and summarized by Cohen *et al.* (1993). The impact of excess sediment pollution along the east coast of Lake Tanganyika, with special reference to the Burundi coast, was assessed for three key groups: ostracodes, fish and diatoms. The following conclusions were reached:

- the three faunal groups showed lower diversities at higher impact sites, although this could be demonstrated statistically only for the ostracodes – a causal relationship between sediment impact and decreasing diversity could not be demonstrated;

- it was suggested that the problem of excess sedimentation should be investigated, starting from a historical record of change through the analysis of the fossil record in sediment cores;

- the importance of size, soil and geological characteristics of the drainage basins are stressed: small drainage basins or those that generate little mud-sized particles in the process of erosion (e.g. quartz sandstone) will have lower risks for impact of sedimentation on biodiversity (e.g. the highly deforested area north and south of Kigoma has experienced little reduction in species richness);

- the following were designated high-risk areas: most of the Burundi coast (already deforested) and much of the Zambian coast (still extensively forested).

Additional data on lake water chemistry and physical properties were reported by Bootsma and Hecky (1993). Martin *et al.* (1993a,b) presented measurements of available oxygen in sediments in various ancient lakes, but without taking degree of impact into account. From their comparative research, they concluded that organisms in Lake Baikal are adapted to much greater sediment oxygen penetration depth than in either of the African lakes. This must be taken into account when assessing hypoxia tolerance in Tanganyika endemics. Excess sedimentation indeed influences various processes related to oxygenation of the sediment–water interface as well as of the sediment itself (see above: influx of nutrients and nutrient dynamics, photosynthetic rates, etc.). Indirectly, oxygen availability may thus be affected by excess sedimentation, and the oxygen requirements of endemic benthic invertebrates must thus be known in natural (low-impact) conditions, in order to assess their reactions to higher impact situations.

As a follow-up to the previous work, a joint international expedition (headed by A. S. Cohen, J.-J. Tiercelin and K. Martens) was organized at the end of 1992. The work was funded by USA and Belgium. Work included census data collected by SCUBA and ROV transects on fish, mollusc and ostracode diversity. Of the latter, densities were also examined below 40 m with remote samplers. Three sites were examined: Cap Banza (Republic of Congo – low-impact site); Luhanga (Republic of Congo – moderate impact) and a site at 28.9 km (northern Burundi – high impact). Several short to medium-long cores were taken at the different sites. Various environmental variables were measured simultaneously. Analysis of the results is still in progress and has thus far been reported by Cohen *et al.* (1993a,b, 1995) and Alin *et al.* (1994). The following preliminary results are presently available:

- 210Pb and photographic analyses have shown that sedimentation rate has increased dramatically over recent years;
- diversity levels are lower at affected sites on rocky bottom for all faunal groups;
- mollusca appear to be less reactive to sediment impacts; ostracodes give statistically significant differences in both species richness and diversity levels between various sites and can also be used to assess historical changes from core analyses;
- trophic analyses suggest that algal grazers may be disproportionately eliminated as a result of turbidity and siltation and the resulting reduction of habitat complexity;
- fossil data indicate that faunal diversity amongst ostracodes at high-impact sites has declined in recent times, with rare taxa disappearing disproportionately;
- highly patchy population structure (both in space and in time) for many key species indicates that dynamic biodiversity assessments, including temporal change in distribution, should replace or at least complement traditional hot-spot analysis of local diversity in reserve design analysis in Lake Tanganyika.

2.11 EFFECT OF INCREASED SEDIMENT LOADS ON FISH

2.11.1 Current level of knowledge

The current level of knowledge on the effects of sediment discharge on the fish community of Lake Tanganyika may be considered as close to nil. However, it is known from other lakes that high discharges of silt and erosion materials affect the fish in rivers and lakes. Although it is reported that most affluents of Lake Tanganyika, and especially those of the northern part, are heavily loaded with sediments in the rainy seasons, only a few specific studies on the effect of this siltation on the fish population have been published (Cohen *et al.*, 1993b,c; Ochi *et al.*, 1993).

The accumulation of sedimentation materials in Lake Tanganyika, and especially in front of the main affluents Malagarasi and Ruzizi, has been analysed several times. However, most of the studies focus on long-term changes or describe global lake-wide accumulation rather than analysing the immediate impact of an increasing sedimentation rate in well-located areas, especially as it bears on fish populations (Ndabigengesere, 1986; Cohen, 1989, 1990, 1994a,b; Coulter, 1991b, 1994; Hecky *et al.*, 1981; Ntakimazi, 1992; Hecky, 1993; Worthington and Lowe-McConnell, 1994).

In recent years, the impact of the northern tributaries of Lake Tanganyika on the fish population has been studied by the UNESCO ecotone project and the CRRHA in close collaboration with the University of Bujumbura (Burundi) and the Centre de Recherche Hydrobiologique (CRH) in Uvira (Republic of Congo). From 1992 to the present, a regular, standardized sampling programme at different locations and different depths, including sites in front of the heavily loaded Mugere and Ntahangwa Rivers, was established. First preliminary results have been reported, but the final results have still to be published (Kimbadi, 1993; Vandelannoote *et al.*, 1993, 1994, 1995; Bigawa *et al.*, 1994; Bitetera, 1994; Niyungeko, 1994; Ntakimazi, 1994, 1995; Risch *et al.*, 1994; Vyumvuhore, 1994; CRRHA, 1995).

2.11.2 Direct impact of sediment discharge on fish

Considering the most important effects of high sedimentation rates on fish populations, four different levels of impact are distinguished.

Fish in closed systems with high levels of suspended materials show a higher incidence of gill diseases, in particular when the particles are abrasive (sand) or fine (clay). In the latter case, silt sticks on the lamellae and interferes with oxygen transfer (Vibert and Lagler, 1961). This phenomenon is mainly observed in smaller fish ponds.

Although very high levels of suspended materials have been measured in some affluents of northern Lake Tanganyika (e.g. in the Mugere and Ntahangwa rivers, especially during rains: Bitetera, 1994; Vandelannoote *et al.*, 1994; Vyumvuhore, 1994), these sediments are carried towards the deeper parts of the lake rather quickly. To date, no effects of direct impact on the fish have been observed. Several species are recorded in silted temporary lagoons in the Ruzizi delta (Kwetuenda, 1983, 1985, 1987).

2.11.3 Changes of environment due to excess sediment discharge

Fish communities in Lake Tanganyika can be heavily disturbed by natural causes, but when this occurs the fish fauna is apparently restored when environmental conditions are normalized. Brichard (1989) stated that after a storm the rocks close to a river mouth were covered with a thin layer of sediment. Typical rock-dwelling cichlids, mainly biocover grazers, disappeared for a while. A couple of days later, when the rocks were freed from sediment by surf action, these fishes returned. On sandy substrate, nearly all fish disappear during jellyfish blooms. After the bloom, the fish community recovers gradually (CRRHA, unpublished data).

In contrast it was reported that community damage caused by sedimentation may have a more severe impact (Cohen *et al.*, 1993b). High deposits of silt in the lake reduce habitat availability by covering up rocks, filling up crevices, etc. Ntakimazi (1995) reported on the silting of the rocky habitat at Gatorongo. Ochi *et al.* (1993) observed the unrecovered disturbance of a fish community on the northern coast of the lake after the renovation of the road. Risch *et al.* (1994) observed that the biodiversity indices (Shannon-Wiener, Menhinick and Margalef) of the fish community on sandy beaches were lowest in front of the mouths of the heavily loaded Ntahangwa and Mugere rivers.

Not only the lake, but also the rivers in the catchment area are affected by sediment discharges (Bizimana and Duchafour, 1991; Niyungeko, 1994; Vandelannoote *et al.*, 1994). De Vos and Snoeks (1994) drew attention to the fact that some riverine populations of non-cichlids are threatened and that, while many affluents of the Tanganyika catchment area appear to be relatively unaffected, other rivers were found to be disturbed by localized erosion.

2.11.4 Impact of excess sediment discharge on the trophic system

Sediment discharges will act upon the phytoplankton, zooplankton and benthic communities (Caljon, 1987, 1991; Hecky, 1993; Wilondja and Cocquyt, 1994) and thus affect the trophic systems of several fish groups. Suspended sediment will reduce visibility which may in turn reduce the foraging success of zooplanktivores and of the early stages of many other fish species.

Deposited sediments will cover the algal carpet on the rocks which is the basis of the littoral food web (Worthington and Lowe-McConnell, 1994). This will primarily affect the specialized aufwuchs eaters, but will also influence other species through interspecific competition, mutualism and commensalism (Hori, 1987). Not all herbivorous species would be affected in the same way. Takamura (1983) described the symbiotic relationship between *Petrochromis polyodon* and *Tropheus* spp., the former removing deposits from the rock surface while grazing. These processed areas were preferred by the latter species for grazing upon the filamentous algae. Hence it can be predicted that *Tropheus* spp. might be much more sensitive to siltation than *Petrochromis* species.

2.11.5 Impact of excess sediment discharge on reproductive behaviour

Deposited sediment might disrupt the behaviour of nest construction of substrate-brooding fish species. A special feature of Lake Tanganyika is that, in contrast to other African lakes, it has a high number of substrate-brooding cichlid species (Coulter, 1991b; Snoeks *et al.*, 1994). This group will probably be under more stress due to high sedimentation deposits than the mouth-brooding species, for the reproduction of which the substratum is less important. Furthermore, a small layer of sediment could, on the micro-scale, diminish the oxygen concentration on the substrates on which eggs and developing embryos are deposited. This phenomenon is known in fish culture but is also reported for an endangered South African cyprinid, *Oreodaimon quathlambae* (Cambray and Meyer, 1988). It is likely to affect non-cichlid species more than cichlids, as all Tanganyikan cichlids show a certain amount of brood care.

2.11.6 Impact of excess sediment discharge on global biodiversity of fish

The ecosystem of Lake Tanganyika is very complex, with a characteristic combination of a high transparency and high primary production (Hecky, 1991; Cohen *et al.*, 1995). Hence it is difficult to comment upon the impact on the global diversity of the lake. A few examples of this complexity are discussed below.

A difficulty with comparative impact studies on fish at different sites around the lake is the occurrence of intralacustrine endemism. The presence or absence of species is not only due to habitat availability, which is an important premise for impact studies, but also to the geological history of the lake. Detailed study of these phenomena has only recently started (Snoeks *et al.*, 1994). At this point it should be stressed that, contrary to general belief, many identification problems persist.

It is most likely that *r*-selected fish groups will be less prone to environmental stress. As most endemic cichlids are *k*-selected strategists, this would mean that this unique group of fish would be threatened by sedimentation effects. On the other hand, it is also known that even specialized cichlid species are potentially able to exploit a wider range of food ('jack-of-all-trades' principle of Liem, 1980; Greenwood, 1984) and that cichlids are morphologically and behaviourally plastic within certain limits (Witte *et al.*, 1990).

2.11.7 Conclusion

Sedimentation appears to be a problem for the fish communities in Lake Tanganyika because of the changes to the environment. Especially rock species could be affected. The impact is highest in the northern part (Cohen *et al.*, 1993). Burundi is the most densely populated area around the lake and the deforestation rate is high. On the other hand, it is difficult to determine whether sedimentation is the principal cause for the apparent lower biodiversity. Another important reason may be the fact that the rocky habitat is much more reduced in the northern part as compared to the south. Also, the impact of the beach seining with mosquito nets on sandy beaches is not insignificant.

Studies of fish communities are rare. Except for the Exploration Hydrobiologique du Lac Tanganyika (1946–47; Van Meel, 1954) and some fisheries-oriented sampling programmes in the 1950s and 1960s (Coulter, 1991a), long-term studies of the dynamics of specific fish communities in the lake began only recently with the Japanese–Zairian project (1977). The continued sampling of CRRHA at specific localities in the northern part since 1992 yielded species that were considered characteristic for the southern part of the lake (*Bathybagrus tetranema*, *Greenwoodochromis christyi*). Preliminary results of the gill-net sampling at different localities and depths are expected to be published soon.

It will be difficult to assess the impact of the sedimentation in one locality if the community structure has not been studied thoroughly.

BIBLIOGRAPHY

ALI, M. M., HAMAD, A. M., SPRINGUEL, I. V. and MURPHY, K. J. (1995) Environmental factors affecting submerged macrophyte communities in regulated waterbodies in Egypt. *Arch. Hydrobiol.*, **133** : 107–128.

ALIN, S., COHEN, A. S., BILLS, R., GASHAGAZA, M. M., MICHEL, E., TIERCELIN, J.-J., MARTENS, K., SOREGHAN, M., COVELIERS, P., WEST, K., NTAKIMAZI, G., MBOKO, S. and KIMBADI, S. (1994) Biodiversity analysis of fish, ostracod and mollusc census data from Lake Tanganyika in relation to sediment disturbance level. In: MÖLSÄ, H. (ed.) *Symposium on Lake Tanganyika Research*, September 1995, Kuopio University, Finland. Abstracts. 34 : 63.

ANON. (1981) *Hydrological Year-Book, 1972–73*. Lusaka: Ministry of Agriculture and Water Development, Republic of Zambia.

ARANUVACHAPUN, S. and WALLING, D. E. (1988) Landsat MSS radiance as a measure of suspended sediment in the lower Yellow river (Hwang Ho). *Remote Sens. Env.*, **25** : 145–165.

ARDEN-CLARKE, C. and EVANS, R. (1993) Soil erosion and conservation in the United Kingdom. pp. 193–215. In: PIMENTAL, D. (ed.) *World Soil Erosion and Conservation*. Cambridge: Cambridge University Press.

ARUDA, J. A., MARZOLF, G. R. and FAULK, R. T. (1983) The role of suspended sediments in the nutrition of zooplankton in turbid reservoirs. *Ecology*, **64** : 1225–1235.

BEAUCHAMP, R. S. A. (1939) Hydrology of Lake Tanganyika. *Int. Rev. Gesamten Hydrobiol.*, **39** : 316–353.

BEAUCHAMP, R. S. A. (1940) Chemistry and hydrography of Lakes Tanganyika and Nyassa. *Nature*, **146** : 253–256.

BERRY, L. and TOWNSHEND, J. (1972) Soil conservation policies in the semi-arid regions of Tanzania, a historical perspective. *Geogr. Annlr.*, **54A** : 241–253.

BIGAWA, S., VANDELANNOOTE, A., OLLEVIER, F. and GRISEZ, L. (1994) L'état de la pollution bactérienne du Lac Tanganyika dans la baie de Bujumbura. pp. 7–8. In: *Résumé des conférences. Journées scientifiques du CRRHA*, 30–31 mars 1994.

BITETERA KANANURA, L. (1994) *Contribution à l'Étude Limnologique des Affluents du Lac Tanganyika. Cas de la Rivière Mugere*. Bujumbura, Burundi: Université du Burundi, Faculté des Sciences.

BIZIMANA, M. and DUCHAFOUR, H. (1991) A drainage basin management study: the case of the Ntahangwa River basin. pp. 43–45. In: COHEN, A. S. (ed.) *Report on the First International Conference on the Conservation and Biodiversity of Lake Tanganyika*, March 1991, Bujumbura, Burundi. Washington, DC: Biodiversity Support Programme.

BOOTSMA, H. A. (1992) Lake Malawi National Park: an overview. pp. 125–128. In: LOWE-McCONNELL, R. H., CRUL, R. C. M. and ROEST, F. C. (eds) *Symposium on Resource Use and Conservation of the African Great Lakes*, Bujumbura, 1989.

BOOTSMA, H. A. and HECKY, R. E. (1993) Conservation of the African Great Lakes: a limnological perspective. *Conserv. Biol.*, **7** : 644–656.

BOOTSMA, H. A. and HECKY, R. E. (in press) Initial measurements of benthic photosynthesis in an African Great Lake. *Hydrobiol.*

BOTZ, R. W. and STOFFERS, P. (1993) Light hydrocarbon gases in Lake Tanganyika hydrothermal fluids (East–Central Africa). *Chem. Geol.*, **104** : 217–224.

BOWEN, S. H. (1984) Evidence of a detritus food chain based on organic precipitates. *Bull. mar. Sci.*, **35** : 440–448.

BRICHARD, P. (1989) *Pierre Brichard's Book of Cichlids and all the other Fishes of Lake Tanganyika*. Neptune City: Tropical Fish Hobbyist Publications.

BROOKS, J. L. and DODSON, S. I. (1965) Predation, body size and the plankton. *Science*, **150** : 28–35.

BRUIJNZEEL, L. A. (1990) *Hydrology of Moist Tropical Forests and Effects of Conversion: a State of Knowledge Review.* Free University of Amsterdam: UNESCO International Hydrological Programme.

BURKHOLDER, J. M. and CUKER, B. E. (1991) Response of periphyton communities to clay and phosphate loading in a shallow reservoir. *J. Phycol.*, **27** : 373–384.

CALJON, A. G. (1987) A recently landlocked brackish-water lagoon of Lake Tanganyika: physical and chemical characteristics and spatio-temporal distribution of phytoplankton. *Hydrobiol.*, **153** : 55–70.

CALJON, A. G. (1991) Sedimentary diatom assemblages in the northern part of Lake Tanganyika. *Hydrobiol.*, **226** : 179–191.

CALJON, A. G. (1992) Water quality in the Bay of Bujumbura (Lake Tanganyika) and its influence on phytoplankton composition. *Mitt. int. Ver. Limnol.* **23** : 55–65.

CALJON, A. G. and COCQUYT, C. Z. (1992) Diatoms from surface sediments of the northern part of Lake Tanganyika. *Hydrobiol.*, **230** : 135–156.

CAMBRAY, J. and MEYER, K. (1988) Early ontogony of an endangered, relict, cold-water cyprinid from Lesotho, *Oreodaimon quathlambae* (Barnard, 1938). *Rev. Hydrobiol. Trop.*, **21** : 309–333.

CAPART, A. (1949) *Sondages et Carte Bathymétrique. Explorations Hydrobiologique du Lac Tanganyika*. Brussels: Institut Royal de Sciences Naturelles de Belgique.

CAPART, A. (1952) *Le Milieu Géographique et Géophysique. Explorations Hydrobiologique du Lac Tanganyika*. Brussels: Institut Royal de Sciences Naturelles de Belgique.

CHAPMAN, L. J., KAUFMAN, L. S., CHAPMAN, C. A. and McKENZIE, F. E. (1995) Hypoxia tolerance in twelve species of East African cichlids: potential for low oxygen refugia in Lake Victoria. *Conserv. Biol.*, **9** : 1274–1288.

COCQUYT, C. (1991) Epilithic diatoms from thrombolitic reefs of Lake Tanganyika. *Belg. J. Bot.*, **124** : 102–108.

COCQUYT, C., CALJON, A. and VYVERMAN, W. (1991) Seasonal and spatial aspects of phytoplankton along the north-eastern coast of Lake Tanganyika. *Ann. Hydrobiol.*, **27** : 215–225.

COCQUYT, C. and JEWSON, D. (1994) *Cymbellonitzschia minima* Hustedt (Bacillariophyceae), a light and electron microscopic study. *Diatom Res.*, **9** : 239–247.

COCQUYT, C. and VYVERMAN, W. (1993) *Surirella sparsipunctata* Hustedt and *S. sparsipunctata* var. *laevis* Hustedt, a light and electron microscopical study. In : *Proceedings of the 12th International Diatom Symposium*, Renesse, Netherlands. *Hydrobiol.*, **269–270** : 97–101.

COCQUYT, C. and VYVERMAN, W. (1994) Composition and diversity of the algal flora in the East African Great Lakes: a comparative survey of Lakes Tanganyika, Malawi (Nyasa) and Victoria. *Arch. Hydrobiol. Beih. Ergebn. Limnol.*, **44** : 161–172.

COCQUYT, C., VYVERMAN, W. and COMPÈRE, P. (1993) A checklist of the algal flora of the East African Great Lakes: Lake Malawi, Lake Tanganyika and Lake Victoria. *Scripta Bot. Belg.*, **8** : 1–56.

COHEN, A. S. (1989) Facies relationships and sedimentation in large rift lakes and implications for hydrocarbon exploration: examples from Lakes Turkana and Tanganyika. *Palaeogeog. Palaeoclimat. Palaeoecol.* **70** : 65–80.

COHEN, A. S. (1990) Tectono-stratigraphic model for sedimentation in Lake Tanganyika, Africa. In : KATZ, B. (ed.) *Lacustrine Basin Expiration – Case Studies and Modern Analogues. Am. Ass. Pet. Geol. Mem.* **50** : 137–150.

COHEN, A. S. (ed.) (1991) *Report on the First International Conference on the Conservation and Biodiversity of Lake Tanganyika*, March 1991, Bujumbura, Burundi. pp. 50–52. Washington, DC: Biodiversity Support Programme.

COHEN, A. S. (1991a) Conservation of the Lake Tanganyika fauna : problems and strategies. pp. 36–38. In: *International Symposium on Limnology and Fisheries of Lake Tanganyika.* Kuopio, Finland: University of Kuopio, Center for Training and Development.

COHEN, A. S. (1992) Criteria for developing viable underwater natural reserves in Lake Tanganyika. pp. 109–116. In: LOWE-McCONNELL, R. H., CRUL, R. C. M. and ROEST, F. C. (eds) *Symposium on Resource Use and Conservation of the African Great Lakes*. Bujumbura, 1989.

COHEN, A. S. (1994a) Extinction in ancient lakes: biodiversity crises and conservation 40 years after J. L. Brooks. pp. 451–479. In: MARTENS, K., GODDEERIS, B. and COULTER, G. (eds) *Speciation in Ancient Lakes. Advances in Limnology* No. 44.

COHEN, A. S. (1994b) Conservation of ancient lake species. pp. 499–500. In: MARTENS, K., GODDEERIS, B. and COULTER, G. (eds) *Speciation in Ancient Lakes. Advances in Limnology* No. 44.

COHEN, A. S., BILLS, R., COCQUYT, C. Z. and CALJON, A. G. (1993) The impact of sediment pollution on biodiversity in Lake Tanganyika. *Conserv. Biol.*, **7** : 667–677.

COHEN, A. S., BILLS, R., GASHAGAZA, M., MICHEL, E., TIERCELIN, J.-J., MARTENS, K., SOREGHAN, M., COVELIERS, P., WEST, K., NTAKIMAZI, G., MBOKO, S. and SONA, K.

(1993a) Sedimentation impacts on biodiversity in Lake Tanganyika, Africa. In: *Abstracts of the SCB/ATB Joint Meeting*, June 1993, Guadalajara, Mexico.

COHEN, A. S., BILLS, R., GASHAGAZA, M., MICHEL, E., TIERCELIN, J.-J., MARTENS, K., SOREGHAN, M., COVELIERS, P., WEST, K., NTAKIMAZIL, G., MBOKO, S. and SONA K. (1993b) Preliminary observations of sedimentation impacts on benthic environments and biodiversity using an ROV submersible in Lake Tanganyika. In: *Abstracts and Programme of PRADI (Poissons, Resources, Diversité)*, Sénégal, November 1993.

COHEN, A. S., SOREGHAN, M. J. and SCHOLZ, C. A. (1993c) Estimating the age of formation of lakes: an example from Lake Tanganyika, East African Rift System. *Geology*, **21** : 511–514.

COHEN, A. S. and THOUIN, C. (1987) Nearshore carbonate deposits in Lake Tanganyika. *Geology*, **15** : 414–418.

COHEN, A. S., WELLS, T., DETTMAN, M. D. and PARK, L. (1995) Paleolimnology and conservation biology problem solving. In: *Abstracts of the 26th Congress of the International Association of Theoretical and Applied Limnology*, 153.

COULTER, G. W. (1963) Hydrological changes in relation to biological production in southern Lake Tanganyika. *Limnol. Oceanogr.*, **8** : 463–477.

COULTER, G. W. (1968) Hydrological processes and primary production in Lake Tanganyika. pp. 609–626. In: *Proceedings of the 11th Conference on Great Lakes Research*. International Association for Great Lakes Research.

COULTER, G. W. (1991a) Composition of the flora and fauna. pp. 200–274. In: COULTER, G. W. (ed.) *Lake Tanganyika and its Life*. London: British Museum (Natural History)/Oxford University Press.

COULTER, G. W. (1991b) Pelagic fish. pp. 111–138. In: COULTER, G. W. (ed.) *Lake Tanganyika and its Life*. London: British Museum (Natural History)/Oxford University Press.

COULTER, G. W. (1994) Lake Tanganyika. pp. 13–18. In: MARTENS, K., GODDEERIS, B. and COULTER, G. (eds) *Speciation in Ancient Lakes. Arch. Hydrobiol. Beih. Ergebn. Limnol.*, **44**.

COULTER, G. W. and SPIGEL, R. H. (1991) Hydrodynamics. pp. 112–134. In: COULTER, G. W. (ed.) *Lake Tanganyika and its Life*. London: British Museum (Natural History)/Oxford University Press.

COVELIERS, P., NTAKIMAZI, G. and ZALIEWSKI, M. (in press) A study on the buffering impact of ecotones on sediment input and nutrient input into Lake Tanganyika.

CRAIG, H. (ed.) (1975) *Lake Tanganyika Geochemical and Hydrographic Study, 1973 Expedition*. La Jolla, California: Scripps Institute of Oceanography.

CRASWELL, E. T. (1993) The management of world soil resources for sustainable agricultural production. pp. 257–275. In: PIMENTAL, D. (ed.) *World Soil Erosion and Conservation*. Cambridge: Cambridge University Press.

CRRHA (1995) *Rapport Annuel 1993–1994*. Bujumbura, Burundi: Centre Régional de Recherche en Hydrobiologie Appliquée.

CUKER, B. E. (1987) Field experiment on the influences of suspended clay and P on the plankton of a small lake. *Limnol. Oceanogr.*, **32** : 840–847.

CUKER, B. E. and HUDSON, L. Jr (1992) Type of suspended clay influences zooplankton response to phosphorus loading. *Limnol. Oceanogr.*, **37** : 566–576.

CULLEN, P. and O'LOUGHLIN, E. M. (1982) *Prediction in Water Quality*. Canberra: Australian Academy of Science.

CUSHING, C. E., MINSHALL, G. W. and NEWBOLD, J. D. (1993) Transport dynamics of fine particulate organic matter in two Idaho streams. *Limnol. Oceanogr.*, **38** : 1101–1115.

DANIDA (1995) *Rapid Water Resources Assessment*, 2 vols. World Bank Report to Ministry of Water Energy and Minerals, United Republic of Tanzania.

DE VOS, L. and SNOEKS, J. (1994) The non-cichlid fishes of the Lake Tanganyika basin. pp. 391–405. In: MARTENS, K., GODDEERIS, B. and COULTER, G. (eds) *Speciation in Ancient Lakes. Arch. Hydrobiol. Beih. Ergebn. Limnol.* No. 44.

DEGENS, E. T. and ITTEKOT, V. (1983) *Dissolved Organic Matter in Lake Tanganyika and Lake Baikal – a Brief Survey*, pp. 129–143. Hamburg: Mitteilungen Geologisch, Paläontologisches Institut, University of Hamburg.

DEGENS, E. T., VON HERZEN, R. P. and WONG, H. K. (1971) Lake Tanganyika: water chemistry, sediments, geological structure. *Naturwissenschaften*, **58** : 229–241.

DICKIE, G. (1880) Notes on algae from lake Nyasa, East Africa. *J. Linn. Soc. Bot.*, **17** : 281–283.

DODSON, S. I. and COOPER, S. D. (1983) Trophic relationships of the freshwater jellyfish *Craspedacusta sowerbyi* Lankester 1880. *Limnol. Oceanogr.*, **28** : 345–351.

DRENNER, R. W., STRICKLER, R. and O'BRIEN, W. J. (1978) Capture probability: the role of zooplankton escape ability in selective feeding of planktivorous fish. *J. Fish. Res. Bd Can.*, **35** : 1370–1373.

DUMONT, H. J. (1986) The Tanganyika sardine in Lake Kivu: another ecodisaster for Africa? *Env. Conserv.*, **13** : 143–148.

DUNNE, T. (1979) Sediment yields and land use in tropical catchments. *J. Hydrol.*, **42** : 281–300.

EDMOND, J. (1975) Lake chemistry. pp. 65–75. In: CRAIG, H. (ed.) *Lake Tanganyika Geochemical and Hydrographic Study, 1973 Expedition.* La Jolla, California: Scripps Institute of Oceanography.

EDMOND, J. M., STALLARD, R. F., CRAIG, H., WEISS, R. F. and COULTER, G. W. (1993) Nutrient chemistry of the water column of Lake Tanganyika. *Limnol. Oceanogr.*, **38** : 725–738.

EDMOND, J., STALLARD, R. F., CRAIG, H., CRAIG, V., WEISS, R. F. and COULTER, G. W. (1995) The chemistry of the water column of Lake Tanganyika, 1. The nutrient elements. *Limnol. Oceaonogr.* (suppl.)

EL-SWAIFY, S. A. (1993) Soil erosion and conservation in the humid tropics. pp. 233–255. In: PIMENTAL, D. (ed.) *World Soil Erosion and Conservation*. Cambridge: Cambridge University Press.

EL-SWAIFY, S. A., DANGLER, E. W. and ARMSTRONG, C. L. (1982) *Soil Erosion by Water in the Tropics*. Research Extension Series No. 24. Hawaii: College of Tropical Agriculture, University of Hawaii.

EVANS, R and SKINNER, R. J. (1987) A survey of water erosion. *Soil Wat.*, **13** : 28–31.

FERGUSON, R. I. and STOTT, T. A. (1987) Forestry effects on suspended sediment and bedload yields in the Balquhidder catchments, Central Scotland. *Trans. R. Soc. Edinb.: Earth Sci.*, **78** : 379–384.

FOURNIER, F. (1960) *Climat et Erosion*. Paris: Presses Universitaires de France.

FOURNIER, F. (1967) Research on soil erosion and soil conservation in Africa. *African Soils*, **12** : 53–96.

FRIEDMAN, M. M. (1980) Comparative morphology and functional significance of copepod receptors and oral structures. pp. 618–622. In: KERFOOT, W. C. (ed.) *Evolution and Ecology of Zooplankton Communities*. Hanover: New England Press.

FRYER, G. (1960) The feeding mechanism of some atyid prawns of the genus *Caridina*. *Trans. R. Soc. Edinb.*, **64** : 217–244.

GARDINER, M. B. (1981) Mechanisms of size-selectivity by planktivorous fish: a test of hypothesis. *Ecology*, **62** : 571–578.

GASSE, F. (1986) *East African Diatoms*. Bibliotheca Diatomologica No. 11.

GASSE, F., LÉDÉE, V., MASSAULT, M. and FONTES, J.-C. (1989) Water-level fluctuations of Lake Tanganyika in phase with oceanic changes during the last glaciation and deglaciation. *Nature*, **342** : 57–59.

GERMAIN, R. (1952). *Les Associations Végétales de la Plaine de la Rusizi (Congo Belge) en Relation avec le Milieu.* Publications de l'Institut National pour l'étude agronomique du Congo belge, Série Scientifique No. 52.

GILLMAN, C. (1933) The hydrology of Lake Tanganyika. *Geol. Surv. Dep. Tanganyika Bull.*, **5**.

GLIWICZ, Z. M. (1977) Food size selection and seasonal succession of filter feeding zooplankton in a eutrophic lake. *Ekol. Pol.*, **25** : 179–225.

GLIWICZ, Z. M. (1986) Suspended clay concentration controlled by filter-feeding zooplankton in a tropical reservoir. *Nature*, **323** : 330–332.

GOURDIN, J., HOLLEBOSH, P., KIBIRITI, C. and NDAYIRAGIJE, S. (1988) *Etude Chimique des Eaux du Lac Tanganyika et ses Affluents au Burundi.*

GRAETZ, R. D. (1991) Desertification: a tale of two feedbacks. In: *Ecosystem Experiments*. Scientific Committee on Problems of the Environment (SCOPE). Chichester: Wiley.

GREENE, C. and JONES, E. N. (1970) *Physical and Chemical Properties of Lake Tanganyika*. Technical memorandum No. 2213-331-70, New London Laboratory, Naval Underwater Systems Center, USA.

GREENWOOD, P. H. (1984) African cichlids and evolutionary theories. pp. 141–154. In: ECHELLE, A. and KORNFIELD, I. (eds) *Evolution of Fish Species Flocks*. University of Maine at Orono Press.

HABERYAN, K. A. (1985) The role of copepod fecal pellets in the deposition of diatoms in Lake Tanganyika. *Limnol. Oceanogr.*, **30** : 1010–1023.

HABERYAN, K. A. and HECKY, R. E. (1987) The late pleistocene and holocene stratigraphy and paleolimnology of lakes Kivu and Tanganyika. *Palaeogeog., Palaeoclimat., Palaeoecol.*, **61** : 169–197.

HANEK, G. (ed.) (1993) *1992 Lake Tanganyika Fisheries Directory*. GCP/RAF/271/FIN-TD/08.

HART, R. C. (1986) Zooplankton abundance, community structure and dynamics in relation to inorganic turbidity, and their implications for a potential fishery in subtropical Lake Le Roux, South Africa. *Freshw. Biol.*, **16** : 351–371.

HART, R. C. (1988) Zooplankton feeding rates in relation to suspended sediment content: potential influences on community structure in a turbid reservoir. *Freshw. Biol.*, **19** : 123–139.

HART, R. C., IRVINE, K. and WAYA, R. (1995) Experimental studies on food dependency of development times and reproductive effort (fecundity and egg size) of *Tropodiaptomus cunningtoni* in relation to its natural distribution in Lake Malawi. *Arch. Hydrobiol.*, **133** : 23–47.

HECKY, R. E. (1981) East African lakes and their trophic efficiencies. In: *Proceedings of AAAS 1981 Meeting, Toronto Symposium.*

HECKY, R. E. (1991) The pelagic ecosystem. pp. 90–110. In: COULTER, G. W. (ed.) *Lake Tanganyika and its Life*. London: British Museum (Natural History)/Oxford University Press.

HECKY, R. E. (1993) The eutrophication of Lake Tanganyika. *Ver. int. Ver. Limnol.*, **25** : 39–48.

HECKY, R. E. and DEGENS, E. T. (1973) Late Pleistocene–Holocene chemical stratigraphy and paleoclimatology in the Rift Valley Lakes of Central Africa. *Technical Report.* Woods Hole Oceanography Institution.

HECKY, R. E. and FEE, E. J. (1981) Primary production and rates of algal growth in Lake Tanganyika. *Limnol. Oceanogr.*, **26** : 532–547.

HECKY, R. E., FEE, E. J., KLING, H. and RUDD, J. W. M. (1978) *Studies on the Planktonic Ecology of Lake Tanganyika.* Technical Report No. 816. Canadian Department of Environmental Fisheries and Marine Service.

HECKY, R. E., FEE, E. J., KLING, H. J. and RUDD, J. W. (1981) *Relationship between Primary Production and Fish Production in Lake Tanganyika. Trans. Am. Fish. Soc.*, **110** : 336–345.

HECKY, R. E. and KLING, H. J. (1981) The phytoplankton and protozooplankton of the euphotic zone of Lake Tanganyika: species composition, biomass, chlorophyll content, and spatio-temporal distribution. *Limnol. Oceanogr.*, **26** : 548–564.

HECKY, R. E. and KLING, H. J. (1987) Phytoplankton ecology of the Great Lakes in the rift valleys of Central Africa. *Arch. Hydrobiol Ergebn. Limnol.,* **25** : 197–228.

HORI, M. (1987) Mutualism and commensalism in a fish community in Lake Tanganyika. pp. 219–239. In: KAWANAO, S., CONNELL, J. and HIDAKA, T. (eds) *Evolution and Coadaptation in Biotic Communities.* University of Tokyo Press.

HORN, W. (1985) Results regarding the food of the planktonic crustaceans *Daphnia hyalina* and *Eudiaptomus gracilis*. *Int. Rev. Ges. Hydrobiol.*, **70** : 702–709

HUC, A. Y., LE FOURNIER, J., VANDENBROUCKE, M., BESSERAU, G., BERNON, M., DA SILVA, M. and FABRE, M. (1989) Northern Lake Tanganyika: a conceptual model of organic sedimentation in a rift lake. In: KATZ, B. and ROSENDAHL, B. R. (eds) *Proceedings of Lacustrine Basin Research Conference*, AAPG, Snowbird, 1988.

HURNI, H. (1993) Land degradation, famine, and land resource scenarios in Ethiopia. pp. 27–61. In: PIMENTAL, D. (ed.) *World Soil Erosion and Conservation*. Cambridge: Cambridge University Press.

HUTCHINSON, G. E. (1930) On the chemical ecology of Lake Tanganyika. *Science*, **71** : 616.

HUTCHINSON, G. E. (1975) *A Treatise on Limnology. Vol. I, Part 1 – Geography and Physics of Lakes*. Chichester: Wiley.

JÓNASSON, P. M., LINDEGAARD, C. and HAMBURGER, K. (1990) Energy budget of Lake Esrom, Denmark. *Verh. int. Ver. Limnol.*, **24** : 632–640.

JACK, J. D., WICKHAM, S. A., TOALSON, S. and GILBERT, J. J. (1993) The effect of clays on a freshwater plankton community: an enclosure experiment. *Arch. Hydrobiol.*, **127** : 257–270.

JACKSON, I. J. (1972) Mean daily rainfall intensity and number of rain days over Tanzania. *Geogr. Annlr.*, **54A**, 369–375.

JOHNSON, R. K., BOSTRÖM, B. and VAN DE BUND, W. (1989) Interactions between *Chironomus plumosus* (L.) and the microbial community in surficial sediments of a shallow eutrophic lake. *Limnol. Oceanogr.*, **34** : 992–1003.

JONES, J. R. and BACHMANN, R. W. (1976) Prediction of phosphorus and chlorophyll levels in lakes. *J. Water Pollut. Control Fed.*, **48** : 2176–2182.

JONES, J. R. and BACHMANN, R. W. (1978) Phosphorus removal by sedimentation in some Iowa reservoirs. *Vehr. int. Ver. Limnol.*, **20** : 1576–1580.

KAWABATA, M. and MIHIGO, N. Y. K. (1982) Littoral fish fauna near Uvira, northwestern end of Lake Tanganyika. *Afr. Study Monogr.*, **2** : 133–143.

KIMBADI, S. (1993) Limnological study of the littoral zone of Lake Tanganyika and rivers flowing into the lake. pp. 86–87. In: NAGOSHI, M., YANAGISAWA, Y. and KAWANABE, H. (eds) *Ecological and Limnological Studies on Lake Tanganyika and its Adjacent Regions*, VIII.

KIRK, K. L. (1991) Inorganic particles alter competition in grazing plankton: the role of selective feeding. *Ecology*, **72** : 915–923.

KIRK, K. L. (1992) Effects of suspended clay on *Daphnia* body growth and fitness. *Freshw. Biol.*, **28** : 103–109.

KIRK, K. L. and GILBERT, J. J. (1990) Suspended clay and the population dynamics of planktonic rotifers and cladocerans. *Ecology*, **71** : 1741–1755.

KNOWLTON, M. F. and JONES, J. J. (1995) Temporal and spatial dynamics of suspended sediments, nutrients and algal biomass in Mark Twain Lake, Missouri. *Arch. Hydrobiol.*, **135** : 145–178.

KOCIOLEK, J. F. and STOERMER, E. F. (1991) New and interesting *Gomphonema* (Bacillariophyceae) species from East Africa. *Proc. Calif. Acad. Sci.*, **47** : 275–288.

KOCIOLEK, J. F. and STOERMER, E. F. (1993) The diatom genus *Gomphocymbella* O. Müller: taxonomy, ultrastructure and phylogenetic relationships. *Nova Hedwigia beiheft* : 106.

KOENINGS, J. P., BURKETT, R. D. and EDMUNSON, M. J. (1990) The exclusion of limnetic cladoceran from turbid glacier-meltwater lakes. *Ecology*, **71** : 57–67.

KUFFERATH, J. (1952) *Le Milieu Biochemique. Résultats Scientifics de l'Exploration Hydrobiologique du Lac Tanganyika (1946–47).* Brussels: Institut des Sciences Naturelles de Belgique, **1** : 31–47.

KUFFERATH, H. (1956) *Organismes Trouvés dans les Carottes de Sondages et les Vases Prélevées au Fond du Lac Tanganyika. Explorations Hydrobiologique du Lac Tanganyika.* No. 4. Brussels: Institut Royal des Sciences Naturelles de Belgique.

KWETUENDA, M. K. (1983) L'importance des lagunes côtières dans le phénomène de reproduction de la faune ichthyologique à l'estuaire de la Ruzizi, Tanganyika. pp. 47–49. In: KAWANABE, H. (ed.) *Ecological and Limnological Study on Lake Tanganyika and its Adjacent Regions*, II.

KWETUENDA, M. K. (1985) Contribution des lagunes côtières dans la reproduction de la faune ichthyologique à l'estuaire de la Ruzizi. pp. 33–37. In: KAWANABE, H. (ed.) *Ecological and Limnological Study on Lake Tanganyika and its Adjacent Regions*, III.

KWETUENDA, M. K. (1987) Fish fauna in temporal lagoons at northern end of Lake Tanganyika. pp. 67–72. In: KAWANABE, H. (ed.) *Ecological and Limnological Study on Lake Tanganyika and its Adjacent Regions*, VI.

LAL, R. (1986) Soil surface management in the tropics for intensive land use and high, sustained production. *Adv. Soil Sci.*, **5** : 1–105.

LAL, R. (1993) Soil erosion and conservation in West Africa. pp. 7–25. In: PIMENTAL, D. (ed.) *World Soil Erosion and Conservation.* Cambridge: Cambridge University Press.

LANGENBERG, V. (1994) Conference on Physico-chemical Studies in the Pelagic Zone of Lake Tanganyika, 30–31 March 1994.

LEHMAN, J. T. (1988) Ecological principles affecting community structure and secondary production by zooplankton in marine and freshwater environments. *Limnol. Oceanogr.*, **33** : 931–945.

LEMA, A. J. (1990) East African climate: 1880–1980. *Water Res. Dev.*, **6** (4).

LEWALLE, J. (1972) Les étages de végétation du Burundi occidental. *Bull. Jard. Nat. Bot. Belg.,* **42** : 1–247.

LIEM, K. F. (1980) Adaptive significance of intra- and interspecific differences in the feeding repertoires of cichlid fishes. *Am. Zool.*, **20** : 295–314.

LIND, O. T. and DÁVALOS-LIND, L. (1992) Association of turbidity and organic carbon with bacterial abundance and cell size in a large, turbid, tropical lake. *Limnol. Oceanogr.*, **36** : 1200–1208.

LIND, O. T., DOYLE, R., VODOPICH, D. S., TROTTER, B. G., LIMÓN, J. G. and DÁVALOS-LIND, L. (1992) Clay turbidity: Regulation of phytoplankton production in a large, nutrient-rich tropical lake. *Limnol. Oceanogr.*, **37** : 549–565.

LOCK, M. A., WALLACE, R. R., COSTERTON, J. W., VENTULLO, R. M. and CHARLTON, S. E. (1984) River epilithon: toward a structural-functional model. *Oikos*, **42** : 10–22.

LOEHNERT, E. P. (1975) *Some Observations on Water Composition and Groundwater Potential in the Lake Tanganyika and Kivu Basins.* Addis Ababa: Economic Commission for Africa.

MAGUIRE, D. J., GOODCHILD, M. and RHIND, D. (1991) *Geographical Information Systems – Principles and Applications*. Harlow, UK: Longman.

MARLIER, G. (1953) Etude biogéographique du bassin de la Ruzizi, basée sur la distribution des poissons. *Ann. Soc. R. zool. Belg.*, **1** : 175–224.

MARLIER, G. (1957) Le Ndagal, poisson pélagique du Lac Tanganyika. *Bull. agric. Congo belge*, **48** : 409–422.

MARTENS, K. (1994) Ostracod speciation in ancient lakes: a review. pp. 203–222. In: MARTENS, K., GODDEERIS, B. and COULTER, G. (eds) *Speciation in Ancient Lakes. Advances in Limnology* No. 44.

MARTIN, P. and BRINKHURST, R. O. (1994) A new species of *Insulodrilus* (Oligochaeta, Phredrilidae) from Lake Tanganyika (East Africa) with notes on the oligochaete fauna of the lake. *Arch. Hydrobiol.*, **130** : 249–256.

MARTIN, P. and GIANI, N. (1995a) *Insulodrilus martensi*, a new species of Phreodrilidae (Oligochaeta) from Lake Tanganyika (East Africa). *Ann. Limnol.*, **31** : 3–8.

MARTIN, P. and GIANI, N. (1995b) Two new species of *Epirodrilus* (Oligochaeta, Tubificidae) from Lake Nyasa and Tanganyika (East Africa), with the redescription of *E. slovenicus* and *E. michaelseni. Zool. Scr.*, **24** : 13–19.

MARTIN, P., GODDEERIS, B. and MARTENS, K. (1993a) Oxygen concentration profiles in soft sediments of Lake Baikal (Russia) near the Selenga Delta. *Freshw. Biol.*, **29** : 343–349.

MARTIN, P., GODDEERIS, B. and MARTENS, K. (1993b) Sediment oxygen distribution in ancient lakes. *Verh. int. Ver. Limnol.*, **25** : 793–794 (extended abstract).

MARTINET, F., JUGET, J. and RIERA, P. (1993) Carbon fluxes across water, sediment and benthos along a gradient of disturbance intensity: adaptive responses of the sediment feeders. *Arch. Hydrobiol.,* **127** : 39–56.

MARZOLF, G. R. (1980) Some aspects of zooplankton existence in surface water impoundments. pp. 1392–1399. In: STEFAN, H. G. (ed.) *Surface Water Impoundments.* American Society for Civil Engineering.

MASIJA, E. H. (1993) Irrigation of wetlands in Tanzania. pp. 73–84. In: KAMUKALA, G. L. and CRAFTER, S. A. (eds) *Wetlands of Tanzania: Proceedings of a Seminar*, Morogoro, Tanzania, 1991, Cambridge, UK: World Conservation Union/IUCN.

MATSUDA, H. and NAMBA, T. (1991) Food web graph of a coevolutionary stable community. *Ecology*, **72** : 267–276.

McCABE, G. D. and O'BRIEN, W. J. (1983) The effects of suspended silt on feeding and reproduction of *Daphnia pulex. Am. Midl. Nat.*, **110** : 324–337.

McCAULEY, E. and DOWNING, J. A. (1985) The prediction of cladoceran grazing rate spectra. *Limnol. Oceanogr.*, **30** : 202–212.

McMANUS, J. and DUCK, R. W. (1988) Internal seiches and subaqueous landforms in lacustrine cohesive sediments. *Nature*, **334** : 511–513.

MELACK, J. M. (1980) An initial measurement of photosynthetic productivity in Lake Tanganyika. *Hydrobiol.*, **72** : 243– 247.

MIHAYO, J. M. (1993) Water supply from wetlands of Tanzania. pp. 67–72. In: KAMULKA, G. L. and CRAFTER, S. A. (eds) *Wetlands of Tanzania: Proceedings of a Seminar*, Morogoro, Tanzania, 1991. Cambridge, UK: World Conservation Union/IUCN.

MILLIMAN, J. D. and SYVITSKY, P. M. (1992) Geomorphic/tectonic control of sediment discharge to the ocean: the importance of small mountainous rivers. *J. Geol.*, **100** : 525–544.

MONDEGUER, A., TIERCELIN, J.-J., HOFFERT, M., LARQUÉ, P., LE FOURNIER, J. and TUCHOLKA, P. (1986) Sédimentation actuelle et récente dans un petit bassin en contexte extensif et décrochant: la Baie de Burton, fossé Nord-Tanganyika, Rift Est-Africaine. Boussens, France: Elf Aquitaine, pp. 229–247.

MPAWENAYO, B. (1985) La flore diatomique des rivières de la plaine de la Rusizi au Burundi. *Bull. Soc. R. Bot. Belg.*, **118** : 141–156.

MPAWENAYO, B. (1986) *De Waters van de Rusizivlakte (Burundi): Milieu, Algenflora Envegetatie.* VUB doctoral thesis, Brussels.

MUZINO, T. (1987) Plankters and benthic organisms in the shore of the north-western part of Lake Tanganyika. pp. 62–63. In: KAWANABE, H. and NAGOSHI, M. (eds) *Ecological and Limnological Study on Lake Tanganyika and its Adjacent Regions*, IV.

NAHIGEZE, A. (1980) *Contribution à l'Étude Phytosociologique de la Végétation des Rivages du Lac Tanganyika*. Thesis, University of Burundi.

NAKAI, K., KAWANABE, H. and GASHAGAZA, M. (1994) Ecological studies on the littoral cichlid communities of Lake Tanganyika: the coexistence of many endemic species. pp. 373–389. In: MARTENS, K., GODDEERIS, B. and COULTER, G. (eds) *Speciation in Ancient Lakes. Arch. Hydrobiol. Beih. Ergebn. Limnol.* No. 44.

NDABIGENGESERE, A. (1986) *La Charge Polluante du Lac Tanganyika par les Arrivées dans la Baie de Bujumbura*. Thèse de doctorat non publié, Université du Burundi, Bujumbura.

NEWMAN, P. and RÖNNBERG, P. (1992) *Changes in Land Utilization within the Last Three Decades in the Babati Area. A Minor Field Study*. Working Paper No. 198. Uppsala, Sweden: Swedish University of Agricultural Sciences.

NEWSON, M. (1992) *Land, Water and Development*. London: Routledge.

NGAMA, J. O. (1992) *Climatic Assessment of Kondoa Eroded Area*. Research Report No. 80. Dar es Salaam: Institute of Resource Assessment, University of Dar es Salaam, Tanzania.

NIYUNGEKO, P. (1994) *Contribution à l'Étude de la Qualité de l'Eau de la Rivière Ntahangwa par les Indicateurs Biologiques Macro-invertébrés*. Mémoire, Université du Burundi, Fac. Sciences.

NSABIMANA, S. (1991) L'érosion des sols et la pollution du Lac Tanganyika au Burundi. pp. 91–94. In: COHEN, A. S. (ed.) *Report on the First International Conference on the Conservation and Biodiversity of Lake Tanganyika*. Washington, DC: Biodiversity Support Program.

NTAKIMAZI, G. (1992) Conservation of the resources of the African Great Lakes: why? An overview. In: LOWE-McCONNELL, R., CRUL, R. and ROEST, F. (eds) *Symposium on Resources and Conservation of the African Great Lakes*, Bujumbura, 1989. *Mitt. int. Ver. Limnol.*, **23** : 5–9.

NTAKIMAZI, G. (1994) Composition et distribution de l'ichtyofaune dans la zone littorale du nord du lac Tanganyika. p. 23. In: *Résumé des Conférences. Journées Scientifiques du CRRHA*, 30–31 mars 1994.

NTAKIMAZI, G. (1995) *Le Rôle des Écotones Terre/Eau dans la Diversité Biologique et les Ressources du Lac Tanganyika*. Rapport final Projet UNESCO/MAB/DANIDA 510/BDI/40, 1991–94. Paris: UNESCO.

OCHI, H., KOHDA, M. and MATSUMOTO, K. (1993) *Ichthyofauna of Permanent Study Area in Northern coast of Lake Tanganyika. Ecological and Limnological Study on Lake Tanganyika and its Adjacent Regions*, VIII.

PAYNE, B. R. (1985) Measurement of rates of accumulation of sediments from radioisotope data. pp. 219–224. In: *Methods of Computing Sedimentation in Lakes and Reservoirs*. UNESCO Technical Report in Hydrology. Paris: UNESCO.

PIMENTAL, D. (1993) Overview. pp. 1–5. In: PIMENTAL, D. (ed.) *World Soil Erosion and Conservation*. Cambridge: Cambridge University Press.

RAPP, A. (1977) *Methods of Soil Erosion Monitoring for Improved Watershed Management*. FAO Conservation Guide No. 1. Rome: FAO.

RAPP, A., AXELSSON, V., BERRY, L. and MURRAY-RUST, D. (1972b) Soil erosion and sediment transport in the Morogoro River Catchment, Tanzania. *Geogr. Annlr.*, **54A** : 125–155.

RAPP, A., BERRY, L. and TEMPLE, P. H. (1972a) Soil erosion and sedimentation in Tanzania – the project. *Geogr. Annlr.*, **54A** : 105–109.

REEKMANS, M., (1985) La végétation de la plaine de la basse Rusizi (Burundi). *Bull. Jard. Nat. Bot. Belg.*, **50** : 401–444.

RIBER, H. and WETZEL, R. G. (1987) Boundary layer and internal diffusion effects on phosphorus fluxes in lake periphyton. *Limnol. Oceanogr.*, **32** : 1181–1194.

RISCH, L., DE VOS, L. and NICAYENZI, F. (1994) Composition et distribution de l'ichtyofaune sur les plages sablonneuses du nord du lac Tanganyika. p. 24. In: *Résumé des Conférences. Journées Scientifiques du CRRHA*, 30–31 mars 1994.

RITCHIE, J. C. and COOPER, C. M. (1988) Comparison of measured suspended sediment concentrations with suspended sediment concentrations measured from Landsat MSS data. *Int. J. Rem. Sens.*, **9** : 379–387.

RODHE, H. and VIRJI, H. (1976) Trends and periodicities in East African rainfall data. *Monthly Weath. Rev.*, **104** : 307–315.

RODIER, J. A. (1983) *Aspects Scientifiques et Techniques de l'Hydrologie des Zones Humides de l'Afrique Centrale* (*Hydrology of Humid Tropical Regions*), pp. 105–126. IAHS Publication No. 140.

ROOSE, E. J. (1977a) Adaptation of soil conservation techniques to the ecological and socio-economic conditions of West Africa. *Agron. Trop.*, **32** : 132–140.

ROOSE, E. J. (1977b) Application of the USLE of Wischmeier and Smith in West Africa. pp. 177–188. In: GREENLAND, D. J. and LAL, R. (eds) *Soil Conservation and Management in the Humid Tropics*. Chichester: Wiley.

ROSENDAHL, B. R., REYNOLDS, D. J., LORBER, P. M., BURGESS, C. F., McGILL, J., SCOTT, D., LAMBIASE, J. J. and DERKSEN, S. J. (1986) Structural expressions of rifting: lessons from Lake Tanganyika, Africa. pp. 29–43. In: FROSTICK, L. E. *et al.* (eds) *Sedimentation in the African Rifts*. Geological Society Special Publication No. 25.

ROSS, R. (1981) Endemism and cosmopolitanism in the diatom flora of East African Great Lakes. pp. 157–177. In: SIMS, R. W., PRICE, J. H. and WHALLEY, P. E. S. (eds) *Evolution, Time and Space: The Emergence of the Biosphere.* London: Academic Press.

SANDSTRÖM, K. (1995) The recent Lake Babati floods in semi-arid Tanzania – a response to changes in land cover? *Geogr. Annlr.*, **77A** : 35–44.

SCHMIDLE, W. (1898) Die von Professor dr. Volkens und Dr. Stuhlmann in Ost-Afrika gesammelten Desmidiaceen. *Bot. Jahrb.*, **26** : 1–59.

SERVICE HYDROLOGIE (1994) *Rapport des Mesures Hydrologiques sur les Rivières Rusizi, Dama, Murembwe, Nyengwe et Rwaba*. Campagne de Septembre 1992 à Octobre 1993. Serv. Hydrol. Inst. Géogr. Burundi. Bujumbura.

SHAPIRO, J. (1973) Blue-green algae: why they become dominant. *Science*, **179** : 382–384.

SNELDER, D. J. and Bryan, R. B. (1995) The use of rainfall simulation tests to assess the influence of vegetation density on soil loss on degraded rangelands in the Baringo District, Kenya. *Catena*, **25** : 105–116.

SNOEKS, J., RÜBER, L. and VERHEYEN, E. (1994) The Tanganyika problem: comments on the taxonomy and distribution patterns of its cichlid fauna. In: MARTENS, K., GODDEERIS, B. and COULTER, G. (eds) *Speciation in Ancient Lakes. Arch. Hydrobiol. Beih. Ergebn. Limnol.*, **44** : 355–372.

STAPPERS, L. (1914) *Composition Chimique de l'Eaux de Surface des Lacs Moëro et Tanganyika.* Brussels: Renseignements de l'Office Colonial belge.

STOCKING, M. (1984) *Rates of Erosion and Sediment Yield in the African Environment.* Challenges in African Hydrology and Water Resources, IAHR Publication No. 144.

STOFFERS, P. and BOTZ, R. (1994) Formation of hydrothermal carbonate in Lake Tanganyika, East-Central Africa. *Chem. Geol.*, **115** : 117–122.

STOFFERS, P. and HECKY, R. E. (1978) Late Pleistocene–Holocene evolution of the Kivu-Tanganyika Basin. *Spec. Pubs Int. Ass. Sedim.*, **2** : 43–55.

STRAKHOV, N. M. (1967) *Principles of Lithogenesis*, Vol. 1. Edinburgh: Oliver and Boyd.

SYMOENS, J. J. (1955a) Observation d'un fleur d'eau à Cyanophycées au Lac Tanganyika. *Fol. sci. Afr. cent.*, **1** : 17.

SYMOENS, J. J. (1955b) Sur le maximum planctonique observé en fin de saison sèche dans le bassin nord du Lac Tanganyika. *Fol. sci. Afr. cent.*, **1** : 12.

SYMOENS, J. J. (1956) Sur la formation de "fleurs d'eau" à Cyanophycées (*Anabaena flos-aquae*) dans le bassin nord du Lac Tanganyika. *Acad. R. Soc. Coloniales, nouvelle série*, **2** : 414–419.

SYMOENS, J. J. (1959) Le développement massif de Cyanophycées planctoniques dans le Lac Tanganyika. In: *Proceedings of the 9th International Botanical Congress, Montreal. Acad. R. Sci. Coloniales*, **24** : 37.

TAKAMURA, K. (1983) Interspecific relationship between two aufwuchs eaters: *Petrochromis polyodon* and *Tropheus moorei* (Pisces: Cichlidae) of Lake Tanganyika, with a discussion on the evolution and functions of a symbiotic relationship. *Physiol. Ecol. Japan*, **20** : 59–69.

TANGHYDRO GROUP (1992) Sublacustrine hydrothermal seeps in northern Lake Tanganyika, East African rift: 1991 Tanganydro expedition, F-31360. Boussens, France: Elf Aquitaine.

TEMPLE, P. H. (1972) Measurements of run-off and soil erosion at an erosion plot scale with particular reference to Tanzania. *Geogr. Annlr.*, **54A** : 203–220.

TIERCELIN, J.-J. and MONDEGUER, A. (1991) The geology of the Tanganyika trough. pp. 7–48. In: COULTER, G. W. (ed.) *Lake Tanganyika and its Life.* London: British Museum (Natural History)/Oxford University Press.

TIETZE, K., GEYH, M., MULLER, H., SCHRODER, L., STAHL, W. and WEHNER, H. (1980) The genesis of the methane in Lake Kivu. *Geologisches Rundschau*, **69** : 452–472.

TIMMS, R. M. and MOSS, B. (1984) Prevention of growth of potentially dense phytoplankton populations by zooplankton grazing, in the presence of zooplanktivorous fish in a shallow wetland ecosystem. *Limnol. Oceanogr.*, **29** : 472–486.

TREFOIS, P. (1994) Monitoring the evolution of desertification processes from 1973 to 1987 in Damagaram (Niger) with Landsat Multi Spectral Scanner and Thematic Mapper. In: *Proceedings SPIE (EUROPTO Series) Conference, Multispectral and Microwave Sensing of Forestry, Hydrology and Natural Resources*, September 1994, Rome, Italy.

TWEDDLE, D. (1994) Conservation and threats to the resources of Lake Malawi. In: LOWE-McCONNELL, R., CRUL, R. and ROEST, F. (eds) *Symposium on Resource and Conservation of the African Great Lakes*, Bujumbura, 1989. *Mitt. int. Ver. Limnol.*, **23** : 17–24.

UNESCO (1978) *World Water Balance and Water Resources of the Earth*. Paris: UNESCO.

UNESCO (1994) *Southern Africa FRIEND Master Register of Gauging Stations*.

VAN MEEL, L. (1954) *Exploration Hydrobiologique du Lac Tanganyika. Le Phytoplancton*. Brussels: Institut Royal des Sciences Naturelles de Belgique. No. 4.

VANDELANNOOTE, A., DEELSTRA, H., VYUMVUHORE, F., BITETERA, L. and OLLEVIER, F. (in press) The impact of the river Ntahangwa, the most polluted Burundian tributary of Lake Tanganyika, on the water quality of the lake.

VANDELANNOOTE, A., VYUMVUHORE, F. and DEELSTRA, H. (1993) Note sur la nécessité de prendre des mesures d'assainissement de la Ntahangwa, l'affluent le plus polluant du bassin nord du Lac Tanganyika. Communication présentée au Workshop *Speciation in Ancient Lakes: Evolution, Biodiversity, Conservation*, Robertville, 1–5 March 1993.

VANDELANNOOTE, A., VYUMVUHORE, F. and BITETERA, L. (1994) L'influence des rivières Ntahangwa et Mugere sur le Lac Tanganyika: les aspects physico-chimiques de la pollution macro-organique et de la pollution physique. p. 4. In: *Résumé des Conférences. Journées Scientifiques du CRRHA*, 30–31 mars 1994.

VERHEYEN, E., BLUST, R. and DECLEIR, W. (1994) Metabolic rate, hypoxia tolerance and aquatic surface respiration of some lacustrine and riverine African cichlid fishes (Pisces, Cichlidiae). *Comp. Biochem. Physiol.*, **107A** : 403–411.

VIBERT, R. and LAGLER, K. F. (1961*) Pêches Continentales. Biologie et Aménagement*. Paris: Dunod.

VINYARD, G. L. and O'BRIEN, W. J. (1976) Effects of light and turbidity on the reactive distance of bluegill (*Lepomis macrochirus*). *J. Fish Res. Bd. Can.*, **33** : 2845–2849.

VYUMVUHORE, F. (1994) *Contribution à l'Analyse de l'État des Pollutions Physique et Organique dans la Rivière Ntahangwa. Mémoire*. Bjumbura: Université du Burundi. Fac. Sciences.

WALLING, D. E. (1984) The sediment yields of African rivers. pp. 265–283. In: *Challenges in African Hydrology and Water Resources*. IAHS Publication No. 144.

WALLING, D. E. and HE, Q. (1993) Towards improved interpretation of ^{137}Cs profiles in lake sediments. pp. 31–53. In: MCMANUS, J. and DUCK, R. W. (eds) *Geomorphology and Sedimentology of Lakes and Reservoirs*. Chichester: Wiley.

WALLING, D. E., PEART, M. R., OLDFIELD, F. and THOMPSON, R. (1979) Suspended sediment sources identified by magnetic measurements. *Nature*, **281** : 110–113.

WARDA (1995) *Rice in West Africa, the Toposequence Approach – Mapping at Multi Levels*, Technical Workshop, Mali, August 1995. M'Be, Côte d'Ivoire: West Africa Rice Development Association.

WEST, G. S. (1907) Report on the freshwater algae of the third Tanganyika expedition. *J. Linn. Soc.*, **38** : 80–197.

WEST, K. and COHEN, A. (1994) Predator–prey coevolution as a model for the unusual morphologies of the crabs and gastropods of Lake Tanganyika. *Arch. Hydrobiol. Beih. Ergebn. Limnol.*, **44** : 267–283.

WETZEL, R. G. (1983) *Limnology*. New York: CBS College Publications.

WETZEL, R. G. (1995) Death, detritus and energy flow in aquatic ecosystems. *Freshw. Biol.* **33** : 83–89.

WILONDJA, K. and COCQUYT, C. (1994) L'étude du phytoplankton dans la zone littorale du lac Tanganyika. pp. 12–13. In: *Résumé des Conférences. Journées Scientifiques du CRRHA*, 30–31 mars 1994.

WISCHMEIER, W. H. and SMITH, D. D. (1978) *Predicting Rainfall Erosion Losses, a Guide to Conservation Planning*. Agriculture Handbook 537. Washington, DC: US Department of Agriculture.

WITTE, F., BAREL, C. D. N. and HOOGERHOUD, R. J. C. (1990) Phenotypic plasticity of anatomical structures and its ecomorphological significance. *Neth. J. Zool.*, **40** (1–2) : 278–298.

WORTHINGTON, E. and LOWE-McCONNELL, R. (1994) African Lakes reviewed: creation and destruction of biodiversity. *Env. Conserv.*, **21** : 199–213.

WOUTERS, K. and MARTENS, K. (1992) Contribution to the knowledge of Tanganyikan cytheraceans, with the description of Mesocyprideis *nom. nov.* (Crustacea, Ostracoda). *Bull. van het Konink. Belgisch. Inst. voor Natuurw. – Biol.*, **62** : 159–166.

WOUTERS, K. and MARTENS, K. (1994) Contribution to the knowledge of the Cyprideis species flock (Crustacea, Ostracoda) of Lake Tanganyika, with the description of three new species. *Bull. van het Konink. Belgisch. Inst. voor Natuurw. – Biol.*, **64** : 111–128.

3. POLLUTION AND ITS EFFECTS ON BIODIVERSITY

3.1 INTRODUCTION

Lake Tanganyika is the largest of the African rift lakes and is also one of the richest waters on earth in terms of species diversity. As yet however, the lake receives little legally mandated environmental protection. The most serious immediate pollution problems threatening the ecosystem result from over-population within the lake basin; major examples are eutrophication (accelerated nutrient enrichment) as well as more traditional forms of pollution involving pesticides and oil residues for example. Together with increased sedimentation and overfishing, particularly in the northern basin, these have led to extinctions of fish species in particular. Not surprisingly, in view of the especially strong links between organisms at different trophic levels in aquatic ecosystems, and the interactions between water chemistry and biota, the losses of fish species can be expected to bring about widespread changes in the ecosystem as a whole.

3.1.1 'Pollution' and 'biodiversity' as applied to Lake Tanganyika

In this chapter, pollution is viewed as the anthropogenically accelerated inputs of the following classes of substances to the lake, via the catchment or directly onto the lake surface in wet or dry deposition, including spillage from ships and boats (see also Alabaster, 1981):

- nutrients, especially phosphorus and nitrogen in catchment run-off and in sewage effluent; these are the main agents of eutrophication *per se* (Harper, 1992) and, more significantly, the cause of enhanced production and biomass accumulation of nuisance plants, e.g. floating hydrophytes such as *Pistia stratiotes*, potentially toxic, planktonic bloom-forming cyanobacteria, and mat-forming algae;

- organic (oxygen-demanding) compounds in sewage, and effluent from, e.g. sugarcane plantations;

- heavy metals from mining and leather tanning industries, etc., often of concern because a number of them accumulate in sediment (Degens and Kulbicki, 1973a), fish food organisms, and thus fish tissue (FAO/SIDA, 1983; Maage *et al.*, 1994);

- pesticides including chlorohydrocarbons stemming from agricultural land including coffee- and cotton-growing areas – residues of many of these compounds also accumulate progressively more acutely in sediments, in sediment-dwelling biota (especially molluscs) and the organisms at the top of the food chains including fish and birds (Deelstra *et al.*, 1976; Matthiessen, 1977);

- materials stemming from oil exploration, including drilling muds (Shackleton, 1980; Livingstone, 1981; Cohen, 1990; Nzori *et al.*, 1990; Baker, 1992; Ndabigengesere, 1992); accidental spills from, e.g. electricity-generating plants, garages;

- miscellaneous substances from, e.g. salt factories.

Gray *et al.* (1991) include 'sewage' and 'litter' (especially plastic products) in addition to the above pollutants, in their manual on marine pollution assessment. It is worth stressing that this review does not consider as 'pollution' the essentially seasonal enrichment of the upper water mass of the lake due to wind-induced upwelling and other hydrological changes (Coulter, 1963, 1967a, 1988). Similarly, we are not concerned here by the emission of sulphurous compounds, hydrocarbons and metals from hydrothermal 'vents' (Degens and Kulbicki, 1973b; Vaslet *et al.*, 1987; Tiercelin, 1988; Tiercelin, Boulegue and Simoneit, 1993; Tiercelin *et al.*, 1989; Tiercelin *et al.*, 1993). Although sedimentation, as a special form of pollution is dealt with separately it may prove difficult in some areas to distinguish between the 'natural' and 'polluted' states as far as eutrophication is concerned. A good example relates to the Ruzizi River: the geology of the upper regions of the catchment of this major feeder of water to Lake Tanganyika ensures that nutrients including phosphate are plentiful long before the river reaches the more heavily populated areas near the lake (Coulter, 1991).

Knowledge about an aquatic ecosystem and the effects of physical and chemical factors on the biota, and the interactions between organisms at different trophic levels, can be furthered significantly by taking account of the sizes, size ranges and size distributions of the organisms (e.g. Gaedke, 1993). This is because size controls a variety of ecological processes (e.g. sinking, vertical migration, light harvesting, nutrient uptake) and interactions (e.g. feeding). Pelagic forms alone are likely to span a range of four or five orders of magnitude in Lake Tanganyika.

It is worth pointing out that localized pollution incidents/situations may well result in more species being recorded. Organisms have a considerable propensity for exploring unusual situations, as exemplified by fish venturing into zones of very low oxygen content (Coulter, 1967b). In providing a 'new' spectra of physical and chemical conditions, hitherto unrecorded species and assemblages can be expected. As an example, the build-up of sediment in and the entry of chemical constituents to the Ruzizi delta (Cohen *et al.*, 1993) may well provide new refuges for invertebrates which could explain the rich avian fauna there (Coulter and Mubamba, 1993). In this connection, one may be faced with the task of judging whether a species assemblage of pollution-associated organisms is more desirable than one of the same numerical richness associated with a more pristine environment. In this connection, there is likely to be a tendency to attach more value to a 'high-profile' fish species forming the basis of a substantial fishery than, say, a key set of microscopic organisms comprising the food chain on which the fish ultimately relies. Different views are also likely to be raised regarding the value of a species where there are conflicts over wildlife or habitat conservation on the one hand, and food production on the other.

3.1.2 Scope of this review

This review assesses current knowledge on the pollution of Lake Tanganyika; as indicated above, it covers eutrophication as well as more traditional aspects of pollution (involving, e.g. metals, herbicides, ship fuel residues). The review results from an examination of some 3000 works. Two works are worthy of mention at the outset: the magnificent monograph on *Lake Tanganyika and its Life* (Coulter, 1991), and a review by Crul (1993). Just over 200 research papers and reports both published and unpublished are cited here, i.e. fewer than 10% of the total number consulted; this reflects the relative paucity of literature on pollution *per se* and of its effects on biodiversity. Thus the bibliography at the end of the chapter can be regarded as being fully comprehensive.

3.2 FEATURES OF LAKE TANGANYIKA THAT GIVE RISE TO CONCERN OVER POLLUTION

3.2.1 Biodiversity

During an expedition some 90 years ago, Cunnington (1913) recorded a total of approximately 320 animals in Lake Tanganyika; these included 146 fish, 84 Mollusca, 31 'Eucopepoda', 22 Ostracoda, 12 macruran Crustacea, six Hydrachnida, five Brachyura, five Polyzoa and three Oligochaeta. Rotifera and Cladoceran Crustacea were not recorded, and the situation regarding these two groups apparently prevails to the present day, although Cladocera have been recorded within the drainage area. (Bearing in mind the apparent ease whereby rotifers can be dispersed, the present authors consider that the likelihood of representatives of this group being completely absent from the lake is slim indeed.) The total number of species in the lake is still not known, but recent inventories suggest that there are more than 1300 animals (Coulter, 1991). Subsequent expeditions show that the number of gastropod molluscs alone now exceeds 130 species (Michel, 1994), while Martens (1994) lists 64 ostracod species, 60 of which are considered to be endemic. Table 3.1 lists animal groups and papers referring to other expeditions, some of which also extend back to the last century (see also Moore, 1897).

Table 3.1 Additional sources of information on species numbers of the better-studied faunal groups (excluding fish) of Lake Tanganyika

Group	Reference/s
Thecamoebae protozoans	Chardez, 1980
Rotifera	Rousselet, 1910
Mollusca	Ancey, 1894; Brown and Mandahl-Barth, 1987; Leloup, 1953
Nematoda	Decremer and Coomans, 1994
Oligochaete annelids	Beddard, 1906
Harpacticoid crustacea	Chappuis, 1955
Branchiuran crustacea	Cunnington, 1913
Ostracod crustacea	Martens, 1984
Copepod crustacea	Leloup, 1952
Aves	Benson and Irwin, 1967; Gaugris, 1979

Fish have attracted most attention in relation to species diversity and speciation (Brooks, 1950; Coulter, 1994b; Martens *et al.*, 1994); some 250 species have been recorded from the lake itself, while a further 100 or so have been found in the basin as a whole. De Vos *et al.* (1994) estimate that nearly 80% of the lacustrine species are endemic. The basin and lake totals, and the numbers of endemic species, are comprised largely of Cichlidae: 187 basin total, 170 lake total and 160 lake endemics. The number of non-cichlid species is still impressive, however, with 72 species in the lake and an additional 70 or so in the basin; 40 of the lacustrine species are thought to be endemic (De Vos and Snoeks, 1994).

Important in exhibiting considerable biodiversity on the plant side are algae. For example, some 750 species have been recorded from the plankton (van Meel, 1954; Hecky and Kling, 1981, 1987; Caljon, 1987). This reflects the spatial variation as well as temporal changes in the assemblages (Cocquyt *et al.*, 1991). However, a feature of the lists of types recorded from the water column is the prominence of species (especially diatoms) which the present authors would associate more with the benthos, i.e. the surfaces of shallow sands, muds and rocks, and fringing macrophytes. This could indicate considerable re-suspension or erosion of material from sediments and other surfaces due to wind-induced turbulence.

As an indication of further ecological diversity, the lake also contains epilithic algae – forms associated more specifically with pebbly, stony and rocky surfaces (Cocquyt, 1991, 1992, 1993). Furthermore, diatoms including subspecies and forms (such as the benthic *Surirella* investigated by Cocquyt and Vyverman, 1992) are likely to be associated with macrophytic vegetation and muds (see e.g. Caljon, 1991). The algal flora of some of the feeder waters, e.g. the Ruzizi (Mpawenayo, 1985, 1986) adds more species to the total list.

In general, biodiversity can be expected to be greater in the littoral areas than the pelagic and the profundal zones (Brooks, 1950). Some investigators (e.g. Dumont, 1994) view the pelagic food webs of the lake as relatively simple; there is only one major, grazing micro-Crustacean, for example, and the apparent absence of Rotifera and Cladocera has been noted elsewhere. High grazing pressure on the algal benthos from a range of fish species alone (see e.g. Hori *et al.*, 1983; Kawanabe, 1981, 1983, 1985, 1989; Kawanabe and Kwetuenda, 1988; Kawanabe and Nagoshi, 1991; Kawanabe *et al.*, 1992) may result in considerable temporal diversity. In this connection, the microphytobenthos may reflect heavy grazing pressure, and thus low biomass in spite of high productivity (see Takamura, 1987), in the same way as the plankton as interpreted by Hecky and Kling (1981, 1987) and Hecky *et al.* (1981). The last-cited paper is also of interest in suggesting the probability that heterotrophic bacterial production exceeds photosynthetic primary production. In the surface waters off Kigoma, potentially bacteriophagous Protozoa were prominent, including colonial vorticellids indicative of organic pollution.

Valuable as they are, many of these records are of somewhat limited use as a baseline for biodiversity with which to compare future findings. This is because studies rarely provide information on the size of samples from which the lists of species were derived.

3.2.2 Factors favouring biodiversity and speciation

The extraordinary species richness of Lake Tanganyika can be attributed to a combination of the following factors:

- the very large size of the lake;
- its great age (>20 million years), and more importantly, isolation for some 1.8 million years;
- a very considerable range in ecological diversity, i.e. extensive physical and chemical spectra providing numerous situations/niches for colonization by organisms; Coulter (1994b), Molsa (1991, 1995) and Nagoshi *et al.*, (1993) are general treatments, while examples concerned with more specific areas are:

  –temporal and spatial changes in light penetration (Plisnier, 1993, 1994), and nutrient concentrations (Edmond, 1975a,b, 1980; Edmond *et al.*, 1993)

  –the variety of rocky, sandy, silty and muddy shorelines (Coenen *et al.*, 1993a; Hori *et al.*, 1983), and land-water ecotones (Ntakimazi, 1995)

  –the interfaces between contrasting conditions of, e.g. oxygen content (Coulter, 1967b), and suspended sediment (Cohen *et al.*, 1993)

  –extremely dynamic and complex water movements (Coulter, 1967a, 1988), and other features of the lake hydrodynamics (Huttula *et al.*, 1993; Huttula and Podsetchine, 1994; Kotilainen, 1993, 1994);

- territorial behaviour of, e.g. cichlids, may also have contributed to the speciation process.

However, some groups of organisms are represented by comparatively restricted species arrays in Lake Tanganyika. Prime examples are the already-noted apparent lack of Rotifera (in open water at least) and of Cladoceran Crustacea; Vuorinen (1993) attributes the absence of Cladocera to the unavailability of oxic sediment for the development of resting stages. The upwelling that brings anoxic water to within 80 m of the lake surface is also suggested by Eccles (1986) to have restricted speciation in demersal fishes.

3.2.3 Features rendering the lake prone and sensitive to pollution

A number of considerations justify concern over the threat of pollution to even such an enormous water body as Lake Tanganyika. Over and above the facts that the lake is very special, there are known hot spots of various forms of pollution (see below), and there is no reason to expect pollution pressures to decrease, there is a need to take into account the following:

- *via* the cascade effect, pollution could bring about changes in the performance and thus the biodiversity of organisms at all trophic levels although, as mentioned above, biodiversity need not necessarily decrease;
- structural and hydrodynamic features of the lake and its basin lend themselves to the accumulation of pollutants and the production of potentially troublesome organisms.

Via the 'trophic cascade' links exist between organisms at different trophic levels in every ecosystem. Realization of this has led to models of trophic interrelationships (see Moreau *et al.*, 1993). The response by one set of organisms, e.g. primary producers, to changes in another, such as the top predators, appears to be particularly well-documented for aquatic systems (e.g. McQueen *et al.*, 1986, 1989). Debate continues over whether a particular system has changed in response to a bottom-up effect, such as a shift in nutrient supplies (eutrophication, for example), or to the top-down effects of some perturbation of fish stocks. There are British examples suggesting strongly that both top fish predation on herbivorous zooplankton and bottom nutrient enrichment are involved (Bailey-Watts and Kirika, 1995). A number of African lakes have been studied in this connection. Lake Kivu (Dumont, 1986) is a case in point, but the most topical example is Lake Victoria. The introduction of the Nile perch (*Lates niloticus*) is considered by most – though not universally (see e.g. Reynolds and Greboval, 1988) – to have played a large part in the degradation of that lake, by way of reducing fish diversity and decimating the artisanal fisheries (Barel *et al.*, 1985; Miller, 1989; Achieng, 1990; Ogutu-Ohwayo, 1990, 1993; Kaufman, 1992; Kudhongania *et al.*, 1992; Kaufman and Cohen, 1993). In this respect, Goldschmidt *et al.* (1993) are justified in concentrating on the cascade effect of the perch. However, the burgeoning of *Eichhornia crassipes* over the last decade at least suggests that eutrophication has also contributed to the situation in Lake Victoria. Figure 3.1 (from Bailey-Watts, 1995) attempts to highlight the latter feature while also acknowledging the impact of *Lates* through its predation on, or competition with, smaller fish species. It is also noteworthy in this connection that another floating hydrophyte, *Pistia stratiotes*, is present in Lake Tanganyka.

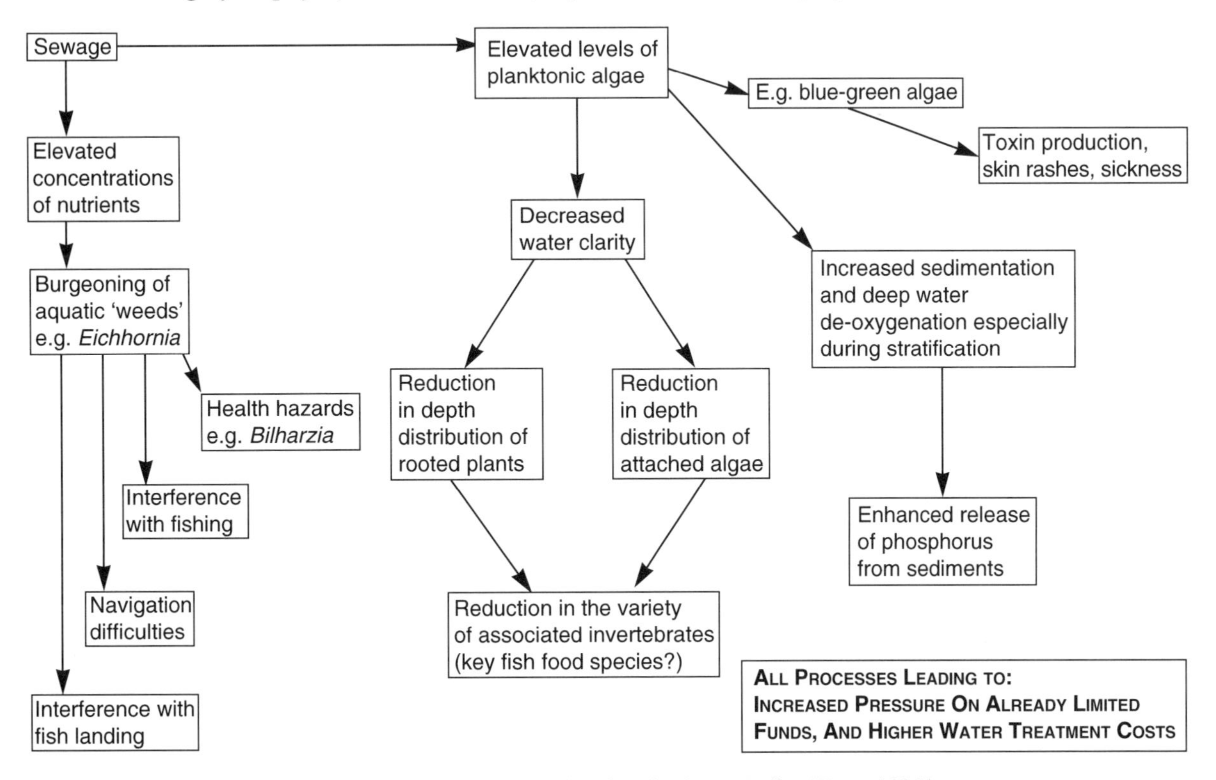

Figure 3.1 Some consequences of eutrophication in Lake Victoria. (From Bailey-Watts, 1995.)

Such cases illustrate very forcibly man's considerable ability to alter some major features of even very large lakes, and in this respect the concern over Lake Tanganyika is well justified. A feature of Lake Tanganyika is its extraordinary dynamism; such a large system might be more usually considered physically well buffered and insensitive to external influences. On the contrary, assemblages of phytoplankton and fish can change remarkably quickly. This can be attributed to fluctuations in factors that impinge on the whole system – examples are incident radiation, wind-induced mixing, massive nutrient upwellings and food-chain interactions. These contrast with impacts such as the external inputs of nutrients and pollutants, which are important in the long term but probably have little significant effect until a threshold level of a particular substance has been attained.

The Lake Victoria situation is very much a reflection of its unique features. At just 40 m mean depth, this African lake is considerably shallower than Lake Tanganyika and thus likely to produce more phytoplankton biomass per unit of (limiting) nutrient loading (Bailey-Watts, 1994, 1995; Bailey-Watts *et al.*, 1994). However, the lessons of Lake Victoria must be noted and, in contrast to the situation surrounding Lake Victoria, early warnings (Fryer, 1972) must be heeded.

Features maximizing the effects of pollution include the virtually completely closed nature of the basin (Coulter, 1991). As a consequence of this and of the prevailing water balance of the lake (Hecky and Bugenyi, 1992), water stays in the lake for an extraordinarily long time (see Table 3.2).

Table 3.2 Hydrological features of Lake Tanganyika resulting in very high retention coefficients for many of the substances that enter the lake

Feature	Value
Annual inflow volume	14 km^3
Annual outflow volume	2.7 km^3
Annual precipitation volume	29 km^3
Flushing time (lake volume/outflow volume)	7000 years
Residence time [lake volume/(precipitation+inflow volume)]	440 years

The system will retain an exceedingly high proportion of the nutrients and pollutants that enter it; only with elements in gaseous phase do major fluxes from the water body take place. For example, models developed in North America and Europe (e.g. Kirchner and Dillon, 1975), in the absence of significant net fluxes of dissolved phosphorus from lake sediments, predict well the retention coefficient of total phosphorus (i.e. that existing in particulate and dissolved form). If morphometric and hydrological data such as those shown in Table 3.2 are incorporated into these formulations, retention coefficients of approximately 0.99 are obtained. Such results are well in keeping with those discussed by Hecky and Bugenyi (1992) and Hecky *et al.* (1994).

3.3 POLLUTION IN LAKE TANGANYIKA

That Coulter (1991) has just two references to 'pollution' in the subject index of his monograph reflects the overwhelming focus on issues other than pollution, although as mentioned above, the vulnerability of Lake Tanganyika to pollution is well recognized. Indeed, it now appears that, in contrast to a quarter of a century ago when Thorslund (1971) concluded that water pollution in African countries was not a serious problem, the situation has changed somewhat. However, there is still dispute over whether Lake Tanganyika is polluted or even whether there are any real threats of pollution.

Firstly, like any freshwater body, Lake Tanganyika acts as a depository of all manner of wind- and water-borne substances from its catchment. In the case of wind-borne pollutants the long-distance transport of contaminants from outside the lake basin and catchment area may also be significant. Secondly, the vital resource – the water itself – is directly affected. Thirdly, the underlying sediment which plays such a key role in the lake ecosystem, especially in the shallower waters, acts both as a repository for and a reservoir of pollutants such as heavy metals and pesticides. Fourthly, fish which, in Africa in general and Lake Tanganyika in particular, constitute a major source of animal protein and vital trace elements (Benemariya *et al.*, 1993), are also likely to become contaminated.

Table 3.3 lists the various types of pollution that could threaten the biodiversity of Lake Tanganyika. It includes relatively easily identified point sources of pollution, as well as the much less readily quantifiable diffuse inputs of pollutants and nutrients, from the catchment and in a number of ways directly over the lake surface. Extreme run-off from steep terrain bordering the east side of the lake, just south of Bujumbura, may also merit attention unless this can be viewed as a natural phenomenon rather than pollution *per se*.

For some pollutants such as urban domestic and industrial wastes, pesticides and heavy metals, a limited amount of information on contamination already exists. However, these data were generated from relatively few sampling sites, and thus provide little information on species, or on seasonal, temporal or spatial variations of the contaminants. To our knowledge, no significant data have been reported for the remaining sources of pollution listed.

Table 3.3 Sources of pollution in the catchment area of Lake Tanganyika

Type of pollution	Sources within lake catchment
Urban domestic wastewater, industrial wastewater	Rivers Ntahangwa, Bujumbura, Burundi[a]
Chlorinated hydrocarbons, pesticides	Cotton plantations in the Ruzizi river[b]
Heavy metals	Northern (Burundi) lake waters[c]
River-borne sediments and associated nutrients from fertilizers	Rivers Ruzizi, Malagarasi
Mercury (inorganic and organic) from gold-mining operations	River Malagarasi catchment
Ash residues	Cement works, Kalemie
Organic wastes, sulphur dioxide	Sugarcane and refining plant, near Uvira
Shipping operations	Lake-wide oil depots, e.g. Bujumbura

[a]Vandelannoote *et al.* (1994; in press); Bigawa *et al.* (1994).
[b]Deelstra (1977a, b)
[c]Sindayigaya *et al.* (1994)

With the ever-increasing pace of agricultural activities in the catchment, and the tendency increasingly to cultivate on steep slopes bordering the lake shores and the inflowing rivers, the amounts of soil being washed into the lake, already significant, are likely to continue to rise. Apart from the physical impact of such materials, the pesticides, nitrates and heavy metals associated with such particulate matter may well constitute a substantial component of the inputs of such contaminants.The lake is also vulnerable to 'informal' industrial activities, e.g. gold mining, which are taking place in the various catchment areas. The scale of such mining operations in developing countries worldwide has expanded rapidly in the last decade, and the use of mercury in gold mining in tropical regions constitutes an increasing and particularly serious source of mercury contamination and exposure (Jernelöv and Ramel, 1995).

Pollution can be as high as 1 kg of mercury for every kg of gold recovered. 70% of such losses are to the atmosphere, with a further 20% escaping as mine tailings discharge to the rivers. Once evaporated, the mercury may remain in the atmosphere for as long as 24 months in dry climates. Reactions in clouds result in shorter overall residence times, causing mercury to return to earth in rainfall. In the Amazon region, for example, such mercury-rich rainwater has been shown to result in pollution of fish collected at sites distant from the mining activities (Velga and Meech, 1995). On the basis of the available literature relating to the use of mercury in gold mining elsewhere in the tropics, the mining of gold in the Malagasi catchment area is likely to have significant environmental implications.

Small-scale diamond mining is also being carried out at various locations in the Lake Tanganyika catchment area. The extent of these operations is unclear at present, but the flotation oils and chemicals used to separate the diamonds from the mineral matrix represent a potential source of pollution. Other industrial facilities, such as the cement works near Kalemie and the sugarcane production and refining operations located north-west of Uvira, may also be sources of pollution. Particulate emissions and high biological oxygen demand (BOD) wastes are likely to represent the most serious environmental threats from the cement works and the sugar production works, respectively.

Shipping activities represent a further potential source of lake pollution. Oils may enter the lake directly from ships due to leakage of fuel oil or from discharged bilge oil. Pollutants may also arise from fuel oil combustion products and from leakage from fuel depots.

3.4 EFFECTS OF POLLUTION ON BIODIVERSITY

Vandelannoote *et al.* (1994; in press) demonstrate the marked contrast between the chemistry of the polluted Ntahangwa and the virtually 'pristine' Mugere in terms of nutrient levels and organic content (for example, 300 and 2 µg soluble reactive phosphorus (as P) per litre, and 923 and 238 mg suspended solids per litre). Those papers comment little on biota, but the phytoplankton chlorophyll *a* levels, which are elevated in the lake bay as a result of the pollution are almost certainly accompanied by shifts in biodiversity; the data of Caljon (1992) support this view, even though no actual biodiversity indices are presented. A similar relationship between increasing pollution levels and decreasing zooplankton and phytoplankton species diversity was observed in the Kenyan waters of Lake Victoria (Foxall *et al.*, 1985). However, in utilizing the potential of macro-invertebrates as indicators of pollution, Vandelannoote *et al.* (1994) suggest that neither the Ntahangwa nor the lake in the vicinity of that inflow is seriously polluted with regard to heavy metals or pesticides. This is in contrast to the situation regarding organic pollution; Bigawa *et al.* (1994) have detected a number of bacteria that indicate faecal contamination of Bujumbura Bay. Although the extent of any effects on biodiversity has not been examined, Bigawa *et al.* (1994) stress the need for more work on the effects of bacterial contamination on other links in the trophic chain. The use of other organisms (e.g. diatoms; see Richardson, 1968) as indicators of pollution needs be explored.

Deelstra (1977a) considered that the pesticide residue levels he measured in fish some 25 years ago were well below the limits set in most African countries at that time. Indeed, he felt that those data would provide a useful benchmark with which future information could be compared and trends thus established. However, Deelstra (1977b) suggests that the organochlorine levels in the fish tissues were higher in Lake Tanganyika than in other African waters. A new census of pesticide purchase and usage could give some indication of the trends, as Deelstra (1977a,b) provides data for the period 1969–72.

Levels of heavy metals measured in two species of fish *(Lates stappersi* and *Stolothrissa tanganyicae)* collected from the Burundi waters of the lake have been reported to be low (Sindayigaya *et al.*, 1994) and were considered by the authors to reflect natural background levels rather than pollution. However, such concentrations may not be representative of the levels from other areas of the lake, or in other fish species.

3.5 GAPS IN CURRENT KNOWLEDGE

That there are serious gaps in knowledge regarding pollution and its effects on biodiversity is already plain from the foregoing sections. This situation is also reflected in the fact that papers and reports on these issues are outnumbered by a factor of *ca* 100:1 by works on other aspects, e.g. lake dynamics, organism ecology and fisheries.

There exist numerous lists of species, but these are commonly restricted to individual phyla or lower taxonomic groups. More importantly, considering the main aim of the present study, data on species composition have invariably been reported without details of the manner in which they were derived.

While a number of pollution hot spots are known, there are virtually no data on the quantities (loadings) of material emanating from them, or how much of the pollutant reaches the lake.

In summary, there appears to be little or no quantitative information on a number of aspects with a strong bearing on pollution and its effects on biodiversity. These are:

- sources of pollution in the catchment and loadings/fluxes to the lake
- as above, but concerning the atmospheric deposition directly onto the lake surface
- the microflora – especially in the littoral benthos
- the zoobenthos in both the littoral shallows and the deeper water sediments/deposits.

Moreover, even where data exist, they have rarely been derived in a manner that allows future investigators to repeat the work; as a result, there are few bodies of information which constitute baselines from which to gauge changes in lake biodiversity.

BIBLIOGRAPHY

ACHIENG, A. P. (1990) The impact of the introduction of Nile perch (*Lates niloticus L.*) on the fisheries of Lake Victoria. *J. Fish Biol.*, **37** (suppl. 1) : 17–23.

ALABASTER, J. S. (1981) *Review of the State of Aquatic Pollution of East African Inland Waters.* United Nations Food and Agriculture Organization Report, CIFA/OP9. Rome: FAO.

ANCEY, C. F. (1894) Sur quelques espèces de mollusques et sur un genre nouveau du Lac Tanganika. *Bull. Soc. zool. France*, **19** : 28–29.

ANCEY, C. F. (1906) Réflexions sur la faune ichthyologique du Lac Tanganika. *Bull. sci. France et Belg.*, **5** : 229.

ANON. (1967–1994) *Statistics for Lake Tanganyika.* Dar es Salaam, Tanzania: Ministry of Tourism, Natural Resources and Environment.

ANON. (1991) *Proceedings and Working Papers, First International Conference on the Conservation and Biodiversity of Lake Tanganyika*, 11–13 March 1991, Bujumbura, Burundi.

BADENHUIZEN, G. R. (1965) *Lufubu River Research Notes.* Fisheries Research Bulletin, 1963–64. Lusaka: Game and Fisheries Department, Ministry of Lands and Natural Resources.

BAILEY-WATTS, A. E. (1986) Seasonal variation in phytoplankton assemblage size spectra in Loch Leven. *Hydrobiol.*, **33** : 25–42.

BAILEY-WATTS, A. E. (1994) Eutrophication. pp. 385–411. In: MAITLAND, P. S., BOON, P. J. and McLUSKY, D. S. (eds) *The Fresh Waters of Scotland: a National Resource of International Significance.* Chichester: Wiley.

BAILEY-WATTS, A. E. (1995) Lake water quality – proposals for strengthening water quality monitoring and for research priorities. In: BULLOCK, A., KEYA, S. O., MUTHURI, F. M., BAILEY-WATTS, A. E., WILLIAMS, R. and WAUGHREY, D. *Lake Victoria Environmental Management Programme: Tasks 11, 16 and 17.* Nairobi, Kenya: FAO.

BAILEY-WATTS, A. E., GUNN, I. D. M. and LYLE, A. A. (in press) Factors affecting the sustainability of Scottish freshwater lochs subject to eutrophication. In: DICKINSON, G. J. (ed.) *Proceedings of a Conference on Sustainability and Resource Management.* Biogeography Research Group, Centre for Research in Environmental Science and Technology/Scottish Natural Heritage.

BAILEY-WATTS, A. E. and KIRIKA, A. (1981) Assessment of size variation in Loch Leven phytoplankton : Methodology and some of its uses in the study of factors influencing size. *J. Plankton Research*, **3** : 261–282.

BAILEY-WATTS, A. E. and KIRIKA, A. (1995) *Phytoplankton Dynamics and the Major Ecological Determinants in Loch Leven NNR during 1994.* Report to Scottish Natural Heritage and the Forth River Purification Board.

BAKER, J. M. (1992) Oil and African Lakes. *Mitt. int. theoret. angew. Limnol.*, **23** : 71–77.

BAREL, C. D. N., DORIT, R., GREENWOOD, P. H., FRYER, G., HUGHES, N., JACKSON, P. B. N., KAWANABE, H., LOWE-McCONNELL, R. H., NAGOSHI, M., RIBBINK, A. L., TREWAVAS, E., WITTE, F. and YAMAOKA, K. (1985). Destruction of fisheries in Africa's lakes. *Nature*, **315** : 19–20.

BAZIGOS, G. P. (1976) *The Design of Fisheries Statistical Surveys. Inland Waters.* FAO Fisheries Technical Paper No. 133. Rome: FAO.

BEDDARD, F. E. (1906) Report on the Oligochaeta, zoological results of the third Tanganyika expedition, 1904–1905. *Proc. zool. Soc. Lond.*, **15** : 206–218.

BENEMARIYA, H., ROBBERECHT, H. and DEELSTRA, H. (1993) Daily dietary intake of copper, zinc and selenium by different population groups in Burundi, Africa. *Sci. Total Env.*, **136** : 49–76.

BENSON, C. W. and IRWIN, M. P. S. (1967) *A Contribution to the Ornithology of Zambia*. Zambia Museum Papers No. 1. Oxford: Oxford University Press.

BIGAWA, S., VANDELANNOOTE, A., OLLEVIER, F. and GRISEZ, L. (1994) L'état de la pollution bactérienne du Lac Tanganyika dans la Baie de Bujumbura. pp. 7–8. In: *Journées Scientifiques du Centre Régional de Recherches en Hydrobiologie Appliquée (CRRHA)* Proceedings of a Conference 30–31 March 1994.

BOOTSMA, H. A. and HECKY, R. E. (1993) Conservation of the African Great Lakes: a limnological perspective. *Conserv. Biol.*, **73** : 644–656.

BROOKS, J. L. (1950) Speciation in ancient lakes. *Quart. Rev. Biol., Baltimore*, **25** : 131–176.

BROWN, D. S. and MANDAHL-BARTH, G. (1987) Living molluscs of Lake Tanganyika, a revised and annotated list. *J. Conchol.*, **32** : 305–327.

BURGIS, M. J. (1984) An estimate of zooplankton biomass for Lake Tanganyika. *Verh. int. Ver. theoret. angew. Limnol., Stuttgart*, **22** : 1199–1203.

BURGIS, M. J. (1986) Food chain efficiency in the open water of Lake Tanganyika. *Bull. séances Acad. r. Sci. d'outre Mer*, **30** : 283–284.

CALJON, A. G. (1987) Phytoplankton of a recently landlocked brackish-water lagoon of Lake Tanganyika: a systematic account. *Hydrobiol.*, **153** : 31–54.

CALJON, A. (1991) Sedimentary diatom assemblages in the northern part of Lake Tanganyika. *Hydrobiol.*, **226** : 179–191.

CALJON, A. G. (1992) Water quality in the Bay of Bujumbura (Lake Tanganyika) and its influence on phytoplankton composition. *Mitt. int. theoret. angew. Limnol.*, **23** : 71–77.

CHAPPUIS, P. A. (1955) Harpacticoïdes psammiques du Lac Tanganika. *Rev. Zool. Bot. afr.*, **51** : 68–80.

CHARDEZ, D. (1980) Sur quelques Thecamoébiens du Lac Tanganika. *Rev. Verv. Hist. Natur.*, **44** : 26–29.

CIFA (1988) *Report on Environmental Activities of the United Nations System in Lake Tanganyika*. CIFA sub-committee for Lake Tanganyika, fourth session, Rome, 25–27 April 1988. DM/LT/87/Inf.4. CIFA.

COCQUYT, C. (1991) Epilithic diatoms from thrombolitic reefs of Lake Tanganyika. *Belg. J. Bot.*, **124** : 102–108.

COCQUYT, C. (1992) Epilithic diatoms from surface sediments of the northern part of Lake Tanganyika. *Hydrobiol.*, **230** : 135–156.

COCQUYT, C. (1993) *Rapport de la Mission au CRRHA: Etude du Plancton et du Benthos du Lac Tanganika, July–August 1993.*

COCQUYT, C., CALJON, A. and VYVERMAN, W. (1991) Seasonal and spatial aspects of phytoplankton along the north-eastern coast of Lake Tanganyika. *Ann. Limnol.*, **273** : 215–225.

COCQUYT, C. and VYVERMAN, W. (1992) *Surirella sparsipunctata* Hustedt and *S. sparsipunctata* var. *laevis* Hustedt. Bacillariophyceae: a light and electron microscopical study. *Hydrobiol.*, **230** : 97–101.

COCQUYT, C. and VYVERMAN, W. (1994) Composition and diversity of the algal flora in the East African Great Lakes: a comparative survey of Lakes Tanganyika, Malawi/Nyassa and Victoria. *Arch. Hydrobiol. Beih. Ergebn. Limnol.*, **44** : 161–172.

COENEN, E. J. (1993a) *Field Guide containing Maps of the Lake Tanganyika Shoreline.* FAO/FINNIDA Research for the Management of the Fisheries on Lake Tanganyika, GCP/RAF/271/FIN-FM/01. Rome: FAO.

COENEN, E. J. (ed.) (1993b) *Historical Data Report on the Fisheries, Fisheries Statistics, Fishing Gears and Water Quality of Lake Tanganyika, Tanzania.* FAO/FINNIDA Research for the Management of the Fisheries on Lake Tanganyika, GCP/RAF/271/FIN-TD/15. Bujumbura, Burundi: FAO.

COENEN, E. J. (1994) *Frame Survey Results for Lake Tanganyika, Burundi, 28–31 October 1992, and Comparison with Past Surveys.* FAO/FINNIDA Research for the Management of the Fisheries on Lake Tanganyika. FAO/FINNIDA Research for the Management of the Fisheries on Lake Tanganyika, GCP/RAF/271/FIN-TD/18. Bujumbura, Burundi: FAO.

COENEN, E. J. (ed.) (1995) *Historical Data Report on the Fisheries Statistics, Limnology, Bromatology, Zooplankton, Fish Biology and Scientific Publications Review of Lake Tanganyika, Zaire.* FAO/FINNIDA, Research for the Management on the Fisheries on Lake Tanganyika. GCP/RAF/271/FIN-TD/32. Bujumbura, Burundi: FAO.

COENEN, E. J., HANEK, G. and KOTILAINEN, P. (1993) Shoreline classification of Lake Tanganyika based on the results of an aerial frame survey, 29 September – 3 October 1992. FAO/FINNIDA Research for the Management of the Fisheries on Lake Tanganyika. GCP/RAF/271/FIN-TN/10. Bujumbura, Burundi: FAO.

COHEN, A. S. (1990) Tectono-stratigraphic model for sedimentation in Lake Tanganyika, Africa. In: KATZ, B. (ed.) Lacustrine basin extirpation – case studies and modern analogues. *Am. Ass. Pet. Geol. Mem.,* **50** : 137– 50.

COHEN, A. S. (1994) Extinction in ancient lakes: biodiversity crises and conservation 40 years after J. L. Brooks. *Arch. Hydrobiol. Beih. Ergehn. Limnol.*, **44** : 451–479.

COHEN, A. S., BILLS, R., COCQUYT, C. and CALJON, A. (1993) The impact of sediment pollution on biodiversity in Lake Tanganyika. *Biol. Conserv.*, **73** : 667– 677.

COMMERFORD, J., EDMOND, J. M., SPIVACK, A. J. and STALLARD, R. F. (1982) The ion balance of Lake Tanganyika and the geochemistry of the Malagarasi River. *Trans. Am. Geophys. Un.*, **63** : 50.

COULTER, G. W. (1963) Hydrological changes in relation to biological production in southern Lake Tanganyika. *Limnol. Oceanogr.*, **8** : 463–477.

COULTER, G. W. (1967a) Hydrological processes in Lake Tanganyika. *Fish. Res. Bull. Zambia*, **4** : 53–56.

COULTER, G. W. (1967b) Low apparent oxygen requirements of deepwater fishes in Lake Tanganyika. *Nature*, **215** : 317–318.

COULTER, G. W. (1968) Hydrological processes and primary production in Lake Tanganyika. pp. 609–626. In: *Proceedings of the 11th Conference on Great Lakes Research.* International Association of Great Lakes Research.

COULTER, G. W. (1988) Seasonal hydrodynamic cycles in Lake Tanganyika. *Verh. int. Ver. theoret. angew. Limnol.*, **23** : 86–89.

COULTER, G. W. (ed) (1991) *Lake Tanganyika and its Life*. London: British Museum (Natural History)/Oxford University Press.

COULTER, G. W. (1994a) Lake Tanganyika. *Arch. Hydrobiol. Beih. Ergebn. Limnol.*, **44** : 13–18.

COULTER, G. W. (1994b) Speciation and fluctuating environments, with reference to ancient East African lakes. *Arch. Hydrobiol. Beih. Ergebn. Limnol.*, **44** : 127–137.

COULTER, G. W. and MUBAMBA, R. (1993) Conservation in Lake Tanganyika, with special reference to underwater parks. *Conserv. Biol.*, **73** : 678–685.

COULTER, G. W. and MUBAMBA, R. (1994) Underwater parks may not be the best conservation tool for Lake Tanganyika – Response. *Conserv. Biol.*, **82** : 330–331.

CRUL, R. C. M. (1993) *Limnology and Hydrology of Lake Tanganyika*. UNESCO/IHP-IV Project M-5-1: Comprehensive and comparative study of great lakes – Monographs of the African Great Lakes. Paris: UNESCO.

CUNNINGTON, W. A. (1913) Report on the Branchiura. Zoological results of the third Tanganyika Expedition (1904–1905). *Proc. zool. Soc. Lond.*, **2** : 262–283.

DECREMER, W. and COOMANS, A. (1994) A compendium of our knowledge of the free-living nematofauna of ancient lakes. *Arch. Hydrobiol. Beih. Ergebn. Limnol.*, **44** : 173–181.

DEELSTRA, H. (1977a) Danger de pollution dans le Lac Tanganika. *Bull. Soc. belge Etud. Géog.*, **46** : 23–35.

DEELSTRA, H. (1977b) Organochlorine insecticide levels in various fish species in Lake Tanganyika. *Med. Fac. Landbouw. Rijksuniv. Gent, Belgium,* **42** : 869–882.

DEELSTRA, H., POWER, J. L. and KENNER, C. T. (1976) Chlorinated hydrocarbon residues in the fish of Lake Tanganyika. *Bull. Env. Contam. Toxicol.*, **15** : 689–698.

DEELSTRA, H. and VAN CAUWENBERGHE, K. (1982) Seasonal changes in fat and fatty acid composition of *Limnothrissa miodon* Boulenger from Lake Tanganyika, with reference to their nutritive value. *Hydrobiol.*, **89** : 123–127.

DEGENS, E. T. and KULBICKI, G. (1973a) *Data File on Metal Distribution in East African Rift Sediments.* Technical Report No. 73-15. Woods Hole, Massachusetts: Woods Hole Oceanographic Institution.

DEGENS, E. T. and KULBICKI, G. (1973b) Hydrothermal origin of metals in some East African Rift lakes. *Mineral. Depos.*, **8** : 388–404.

DEGENS, E. T., VON HERZEN, R. and WONG, H.-K. (1971) Lake Tanganyika: water chemistry, sediments, geological structure. *Naturwissenschaften*, **585** : 229–241.

DE VOS, L., SEGERS, L., TAVERNE, L. and VAN DEN AUDENAERDE, T. (1994) *Composition et Distribution de l'Ichtyofaune dans les Affluentes du Nord du Lac Tanganika*. Rapport sur les deuxièmes Journées Scientifiques du Centre Régional de Recherche en Hydrobiologie Appliquée, 30–31 mars, 1994. Bujumbura, Burundi: CRRHA.

DE VOS, L. and SNOEKS, K. (1994) The non–cichlid fishes of the Lake Tanganyika basin. *Arch. Hydrobiol. Beih. Ergebn. Limnol.*, **44** : 391–405.

DUBOIS, J. Th. (1958a) Evolution de la température, de l'oxygène dissous et de la transparence dans la baie nord du Lac Tanganika. *Hydrobiol.*, **10** : 215–240.

DUBOIS, J. Th. (1958b) Composition chimique des affluents du nord du Lac Tanganika. *Bull. séances Acad. r. Sci. d'outre Mer*, **4** : 1226–1237.

DUMONT, H. J. (1986) The Tanganyika sardine in Lake Kivu: another ecodisaster for Africa? *Environ. Conserv.*, **132** : 143–148.

DUMONT, H. J. (1994) Ancient lakes have simplified pelagic food webs. *Arch. Hydrobiol. Beih. Ergebn. Limnol.*, **44** : 223–234.

ECCLES, D. H. (1986) Is speciation of demersal fishes in Lake Tanganyika restrained by physical limnological conditions? *Biol. J. Linn. Soc.*, **29** : 115–122.

EDMOND, J. (1975a) *Report on a Study of Nutrient Chemistry, Lake Tanganyika, March–April 1975.* FAO Report FI:DP/BDI/70/508. Rome: FAO.

EDMOND, J. (1975b) Lake chemistry. pp. 65–75. In: CRAIG, H. (ed.) *LakeTanganyika Geochemical and Hydrographic Study, 1973 Expedition.* Series 75, No. 5. La Jolla, California: Scripps Institute of Oceanography.

EDMOND, J. (1980) Chemistry of Lake Tanganyika, a 1400 m, thermally stratified, rift valley lake. *Trans. Am. Geophys. Un.*, **61** : 1004.

EDMOND, J., STALLARD, R. F., CRAIG, H., CRAIG, V., WEISS, R. F. and COULTER, G. W. (1993) Nutrient chemistry of the water column of Lake Tanganyika. *Limnol. Oceanogr.*, **38** : 725–738.

FAO/SIDA (1983) *Manual of Methods in Aquatic Environmental Research . Part 9. Analyses of Metals and Organochlorines in Fish.* Fisheries Technical Paper No. 212. Rome: FAO.

FERRO, W. (1975) *Data Files, Limnology – Lake Tanganyika Northern Bay, Data Collected 1972–1975.* FAO Report FI:DP/BDI/73/020. Rome: FAO.

FERRO, W. and COULTER, G. W. (1974) *Limnological Data from the North of Lake Tanganyika.* FAO Report F1:DP/BDI/73/020/10. Rome: FAO.

FOXALL, C. D., LITTERICK, M. and NJUGUNA, S. (1985) *Report on the Baseline Study of the Water Quality of Lake Victoria, Kenya*, Vol. I. Kisumu, Kenya: Lake Basin Development Authority.

FRYER, G. (1972) Conservation of the Great Lakes of Africa: a lesson and a warning. *Biol. Conserv.*, **4** : 256–262.

GAEDKE, U. (1993) Ecosystem analysis based on biomass size distributions: a case study of a plankton community in a large lake. *Limnol. Oceanogr.*, **381** : 112–127.

GAUGRIS, Y. (1976) Additions à l'inventaire des oiseaux du Burundi. *L'Oiseau Rev. Française d'Ornithol.*, **46** : 273–289.

GAUGRIS, Y. (1979) Les oiseaux aquatiques de la plaine de la basse Rusizi (Burundi). *L'Oiseau Rev. Française d'Ornithol.*, **49** : 133–153.

GEF (1994). *Biodiversity: Definition by the Convention on Biological Diversity.* Global Environment Facility.

GOLDSCHMIDT, T., WITTE, F. and WANINK, J. (1993) Cascading effects of the introduced Nile perch on detrivorous/planctivorous species in the sublittoral areas of Lake Victoria. *Conserv. Biol.*, **73** : 686–700.

GRAY, J. S., McINTYRE, A. D. and STIRN, J. (1991) *Manual of Methods in Aquatic Environment Research, Part 11. Biological Assessment of Marine Pollution with Particular Reference to Benthos.* Rome: FAO.

GREENE, C. and JONES, E. N. (1970) *Physical and Chemical Properties of Lake Tanganyika.* Technical Memorandum No. 2213-331-70, New London Laboratory Naval Underwater Systems Center, USA.

HANEK, G. (ed.) (1993) *Lake Tanganyika Fisheries Directory.* FAO/FINNIDA Research for the Management of the Fisheries on Lake Tanganyika, GCP/RAF/271/FIN-TD/08. Bujumbura, Burundi: FAO.

HANEK, G. (1994) *Management of Lake Tanganyika Fisheries Resources.* FAO/FINNIDA Research for the Management of the Fisheries on Lake Tanganyika, GCP/RAF/271/FIN-TD/25. Bujumbura, Burundi: FAO.

HANEK, G. (ed.) (1995) *R/V Tanganyika Explorer: Guidelines.* FAO/FINNIDA Research for the Management of the Fisheries on Lake Tanganyika, GCP/RAF/271/FIN-FM/15. Bujumbura, Burundi: FAO.

HANEK, G. and COENEN, E. J. (1994) *Report on LTR's Second Scientific Sampling Programme Assessment Meeting*, Kigoma, 11–12 April 1994. FAO/FINNIDA Research for the Management of the Fisheries on Lake Tanganyika, GCP/RAF/27L/FIN-TD/17. Bujumbura, Burundi: FAO.

HANEK, G., COENEN, E. J. and KOTILAINEN, P. (1993) *Aerial Frame Survey of Lake Tanganyika Fisheries.* FAO/FINNIDA Research for the Management of the Fisheries on Lake Tanganyika, GCP/RAF/271/FIN-TD/09. Bujumbura, Burundi: FAO.

HARPER, D. M. (1992) *Eutrophication of Freshwaters: Principles, Problems and Restoration.* London: Chapman & Hall.

HECKY, R. E. (1984) African lakes and their trophic efficiencies: a temporal perspective. pp. 405–448. In: MEYERS, D. G. and STRICKLER, J. R. (eds) *Trophic Interactions Within Aquatic Ecosystems.* AAAS Symposium No. 85. Washington, DC: Westview Press.

HECKY, R. E. (1993) The eutrophication of Lake Victoria. *Verh. int. Ver. theoret. angew. Limnol., Stuttgart*, **25** : 39–48.

HECKY, R. E., BOOTSMA, H. A., MUGIDDE, R. and BUGENYI, F. W. B. (1994) Phosphorus pumps, nitrogen sinks and silicon drains: plumbing nutrients in the African Great Lakes. In: *The Limnology, Climatology and Palaeoclimatology of the East African Lakes.* International Symposium sponsored by the International Decade for the East African Lakes (IDEAL), Jinja, Uganda, February 1993.

HECKY, R. E. and BUGENYI, F. W. B. (1992). Hydrology and chemistry of the African Great Lakes and water quality issues: problems and solutions. *Mitt. int. Ver. theoret. angew. Limnol.*, **25** : 39–48.

HECKY, R. E., FEE, E. J., KLING, H. J. and RUDD, J. W. (1978) *Studies on the Planktonic Ecology of Lake Tanganyika.* Fisheries and Marine Service Technical Report No. 816. Canadian Department of Fish and Environment.

HECKY, R. E., FEE, E. J., KLING, H. J. and RUDD J. W. (1981) Relationship between primary production and fish production in Lake Tanganyika. *Trans. Am. Fish. Soc.*, **110** : 336–345.

HECKY, R. E. and KLING, H. J. (1981) The phytoplankton and protozooplankton of the euphotic zone of Lake Tanganyika: species composition, biomass, chlorophyll content, and spatio-temporal distribution. *Limnol. Oceanogr.*, **26** : 548–564.

HECKY, R. E. and KLING, H. J. (1987) Phytoplankton ecology of the Great Lakes in the rift valleys of Central Africa. *Arch. Hydrobiol. Beih. Ergebn. Limnol.,* **25** : 197–228.

HENDERSON, H. F. and WELCOMME, R. L. (1974) *The Relationship of Yield to Morpho-edaphic Index and Numbers of Fishermen in African Inland Fisheries.* FAO/CIFA Occasional Papers No. 1. Rome: FAO.

HORI, M., YAMAOKA, K. and TAKAMURA, K. (1983) Abundance and micro-distribution of cichlid fishes on a rocky shore of Lake Tanganyika. *Afr. Study Monogr., Kyoto Univ.*, **3** : 25–38.

HUC, A. Y., LEFOURNIER, J., VANDENBROUCKE, M. and BESSEREAU, G. (1990) Northern Lake Tanganyika; an example of organic sedimentation in an anoxic rift lake. In: KATZ, B. J. (ed.) *Lacustrine Basin Exploration: Case Studies and Modern Analogs.* AAPG Memoirs No. 50.

HUTCHINSON, G. E. (1930). On the chemical ecology of Lake Tanganyika. *Science*, **71** : 616.

HUTTULA, T., PELTONEN, A. and NIEMINEN, J. (1993) *Hydrodynamic Measurements on Lake Tanganyika.* FAO/FINNIDA Research for the Management of the Fisheries on Lake Tanganyika, GCP/RAF/271/FIN-FM/02. Bujumbura, Burundi: FAO.

HUTTULA, T. and PODSETCHINE, V. (1994) *Modèle Hydrologique au Lac Tanganyika.* FAO/FINNIDA Recherche pour l'Aménagement des Pêches au Lac Tanganyika, GCP/RAF/271/FIN-TD/20. Bujumbura, Burundi: FAO.

IoH (1993) *HYQUAL Water Quality Database Operation Manual.* Wallingford, UK: Institute of Hydrology.

JERNELÖV, A. and RAMEL, C. (1995) Evaluation of the role and distribution of mercury in ecosystems with special emphasis on tropical regions. *Ambio*, **24** : 319–320.

KAUFMAN, L. (1992) Catastrophic change in species-rich freshwater ecosystems. The lessons of Lake Victoria. *BioScience*, **42** : 846–858.

KAUFMAN, L. and COHEN, A. (1993). The Great Lakes of Africa. *Conserv. Biol.*, **73** : 632–633.

KAWANABE, H. (ed.) (1981) *Ecological and Limnological Study on Lake Tanganyika and its Adjacent Regions*, I. Kyoto, Japan: Kyoto University.

KAWANABE, H. (ed.) (1983) *Ecological and Limnological Study on Lake Tanganyika and its Adjacent Regions*, II. Kyoto, Japan: Kyoto University.

KAWANABE, H. (ed.) (1985) *Ecological and Limnological Study on Lake Tanganyika and its Adjacent Regions*, III. Kyoto, Japan: Kyoto University.

KAWANABE, H. (ed.) (1988) *Ecological and Limnological Study on Lake Tanganyika and its Adjacent Regions*, V. Kyoto, Japan: Kyoto University.

KAWANABE, H. (ed.) (1989) *Ecological and Limnological Study on Lake Tanganyika and its Adjacent Regions*, VI. Kyoto, Japan: Kyoto University.

KAWANABE, H. and NAGOSHI, M. (eds) (1991) *Ecological and Limnological Study on Lake Tanganyika and its Adjacent Regions*, IV. Kyoto, Japan: Kyoto University.

KAWANABE, H., KWETUENDA, M. K. and GASHAGAZA, M. M. (1992) Ecological and limnological studies of Lake Tanganyika and its adjacent regions between African and Japanese scientists: an introduction. *Mitt. int. Ver. theoret. angew. Limnol.*, **232** : 79–84.

KILHAM, P. and HECKY, R. E. (1973) Fluoride. Geochemical and ecological significance in East African waters and sediments. *Limnol. Oceanogr.*, **18** : 932–945.

KIRCHNER, W. B. and DILLON, P. J. (1975) An empirical method of estimating the retention of phosphorus in lakes. *Water Resourc. Res.*, **11** : 182–183.

KONDO, T. and HORI, M. (1986) Abundance of zooplankters on a rocky shore of Lake Tanganyika: a preliminary report. *Afr. Stud. Monogr.*, **6** : 17–23.

KOTILAINEN, P. (1993) *Field Notes for Hydrodynamic Studies.* FAO/FINNIDA Research for the Management of the Fisheries on Lake Tanganyika, GCP/RAF/271/FIN-FM/10. Bujumbura, Burundi: FAO.

KOTILAINEN, P. (1994) *Field Notes for Hydrodynamic Studies.* FAO/FINNIDA Research for the Management of the Fisheries on Lake Tanganyika, GCP/RAF/271/FIN-FM/10. Bujumbura, Burundi: FAO.

KUDHONGANIA, A. W., TWONGO, T. and OGUTO-OHWAYO, R. (1992) Impact of the Nile perch on the fisheries of Lake Victoria and Kyoga. *Hydrobiol.*, **232** : 1–10.

KURKI, H. (1993) *Field Notes on Zooplankton.* FAO/FINNIDA Research for the Management of the Fisheries on Lake Tanganyika, GCP/RAF/271/FIN-FM/09. Bujumbura, Burundi: FAO.

KUWAMURA, T. (1986a) Substratum spawning and biparental guarding of the Tanganyikan cichlid *Boulengerochromis microlepis* with notes on its life history. *Physiol. Ecol. Japan*, **23** : 31–43.

LELOUP, E. (1952) Les invertébrés. Résultats scientifiques de l'exploration hydrobiologique du Lac Tanganika (1946–1947). *Inst. R. Sci. Nat. Belgique*, **1** : 71–100.

LELOUP, E. (1953) Gastéropodes. Résultats scientifiques de l'exploration hydrobiologique du Lac Tanganika (1946–(1947). *Inst. R. Sci. Nat. Belgique*, **3** : 1–272.

LINDQVIST, O. V. and MIKKOLA H. (1989) *Lake Tanganyika: Review of Limnology, Stock Assessment, Biology of Fishes and Fisheries.* FAO Report GCP/RAF/229/FIN. Rome: FAO.

LIVINGSTONE, D. A. (1981) Deep drilling in African lakes. *Palaeoecol. Africa Sur. Isl. Antarc.*, **13** : 121.

LOWE-McCONNELL, R. H. (1956) The breeding behaviour of *Tilapia* species (Pisces: Cichlidae) in natural waters: observations on *T. karomo* Poll and *T. variabilis* Boulenger. *Behaviour*, **9** : 140–163.

MAAGE, A., ECKHOFF, K. and KJELLEVOLD, M. (1994) *Fluorine, Iodine, Iron, Zinc and Selected Fatty Acid Profiles in Fish and Staple Food from East Africa.* FAO project GCP/INT/467/NOR, Fish in Nutrition. Bergen, Norway: Institute of Nutrition, Directorate of Fisheries.

MADDEN, C. J. and DAY, J. W. (1992) An instrument system for high-speed mapping of chlorophyll *a* and physico-chemical variables in surface waters. *Estuaries*, **15** : 421–427.

MANN, K. H. (1993) Physical oceanography, food chains, and fish stocks: a review. *ICES J. mar. Sci.*, **50** : 105–119.

MARTENS, K. (1984) Annotated checklist of non-marine ostracods (Crustacea: Ostracoda) from African inland waters. *Doc. zool. Mus. r. Afr. cent.*, **20** : 1–51.

MARTENS, K. (1994) Ostracods in ancient lakes. *Arch. Hydrobiol. Beih. Ergebn. Limnol.*, **44** : 203–222.

MARTENS, K., GODDEERIS, B. and COULTER, G. (eds) (1994) Speciation in ancient lakes: advances in limnology. *Arch. Hydrobiol. Beih. Ergebn. Limnol.*, **44**, 1–508.

MATTHIESSEN, P. (1977) A visit to Tanzania with reference to the problem of pesticide run-off into Lake Tanganyika. *Limnologie*, **22** : 26–62.

McQUEEN, D. J., POST, J. R. and MILLS, E. L. (1986) Trophic relationships of freshwater pelagic ecosystems. *Can. J. Fish. aquat. Sci.*, **43** : 1571–1581.

McQUEEN, D. J., JOHANNES, M. R. S., POST, J. R., STEWART, T. J. and LEAN, D. R. S. (1989) Bottom-up and top-down impacts on freshwater pelagic community structure. *Ecol. Monogr.*, **59** : 289–309.

MEEL, L. VAN (1954) *Le Phytoplancton – Résultats Scientifiques de l'Exploration Hydrobiologique du Lac Tanganika, 1946–1947*. Brussels: Institut Royal des Sciences Naturelles de Belgique.

MELACK, J. M. (1976) Primary productivity and fish yields in tropical lakes. *Trans. Am. Fish. Soc.*, **105** : 575–580.

MEYBECK, M. (1985) *Evaluation Préliminaire de la Pollution du Lac Tanganyika.* Nairobi, Kenya: UNESCO.

MICHEL, E. (1994) Why snails radiate: a review of gastropod evolution in long-lived lakes, both recent and fossil. *Arch. Hydrobiol. Beih. Ergebn. Limnol.*, **44** : 285–317.

MILLER, D. J. (1989) Introductions and extinctions of fish in the African Great Lakes. *Trends Ecol. Evol.*, **42** : 56–59.

MOLSA, H. (1991) *Proceedings of an International Symposium on Limnology and Fisheries of Lake Tanganyika*. Kuopio, Finland: University of Kuopio, Center for Training and Development.

MOLSA, H. (1995) *Symposium on Lake Tanganyika Research, 11–15 September 1995, Kuopio, Finland.* Abstracts. p.117.

MOORE, J. E. S. (1897) On the general zoological results of the Tanganyika Expedition. *Proc. zool. Soc. Lond.*, 436–439.

MOREAU, J., NYAKACYENI, B., PEARCE, M. and PETIT, P. (1993) *Trophic Relationships in the Pelagic Zone of Lake Tanganyika, Burundi Sector*. Toulouse, France/Bujumbura, Burundi: Laboratoire d'Icthyologie Appliquée.

MPAWENAYO, B. (1985) La flore diatomique des rivières de la plaine de la Rusizi au Burundi. *Bull. Soc. R. Bot. Belgique*, **118** : 141–156.

MPAWENAYO, B. (1986) *De Waters van de Ruzizi-vlakte (Burundi): Milieu, Algenflora en-vegetatie*. PhD thesis, Vrije Universiteit, Brussels.

NAGOSHI, M., YANAGISAWA, Y. and KAWANABE, H. (eds) (1993) *Ecological and Limnological Studies on Lake Tanganyika and its Adjacent Regions*, VIII. Kyoto, Japan: Kyoto University.

NARITA, T. (1983) Species composition, vertical distribution and density of zooplankters, and some limnological features off the coast of Mahale Mountains in Lake Tanganyika. pp. 12–14. In: KAWANABE, H. (ed.) *Ecological and Limnological Study of Lake Tanganyika*, Vol. 2. Kyoto: Kyoto University Press.

NARITA, T., MULIMBWA, N. and MIZUNO, T. (1985) Vertical distribution and seasonal abundance of zooplankters in Lake Tanganyika. *Afr. Stud. Monogr. Kyoto*, **6** : 1–16.

NDABIGENGESERE, A. (1992) *La Charge Polluante du lac Tanganika par les Arrivées dans la Baie de Bujumbura*. Bujumbura, Burundi: University of Burundi.

NDAHIGEZE, S. (1995) *Updated Catalogue of Regional Documentation Centre for Lake Tanganika Fisheries Research*. FAO/FINNIDA Report GCP/RAF/271/FIN-FM/16. Rome: FAO.

NDAYIZEYE, P. (1985) *Influence des Déchets Industriels sur la Qualité de l'Eau du Lac Tanganika*. Mémoire de Maîtrise en Biologie. Bujumbura, Burundi: Université du Burundi.

NTAKIMAZI, G. (1995) *Le Rôle des Écotones Terre/Eau dans la Diversité Biologique et les Ressources du Lac Tanganika*. Projet UNESCO/MAB/DANIDA 510/BDI/40, 1991–1994. Paris: UNESCO.

NZORI, S., TARANYENKO, Y. I., MPANDA, S. and MBEDE, E. (1990) *Petroleum Prospects of the East African Rift Lake-basins of Tanganyika, Nyasa and Rukwa. Proceedings of 15th Colloquium of African Geology*, Nancy, September 1990. Occasional Publication No. 20, International Center for Training and Exchanges in the Geosciences.

OGUTU-OHWAYO, R. (1990) The reduction in fish species diversity in lakes Victoria and Kyoga, East Africa, following human exploitation and introduction of non-native fishes. *J. Fish Biol.*, **37** (suppl. A) : 207–208.

OGUTU-OHWAYO, R. (1993) The effects of predation by Nile perch, *Lates niloticus* L., on the fish of Lake Nabugabo, with suggestions for conservation of endangered endemic cichlids. *Conserv. Biol.*, **73** : 70l–711.

PLISNIER, P. D. (1993) *Field Manual for Limnological Sampling on Lake Tanganyika*. FAO/FINNIDA Research for the Management of the Fisheries on Lake Tanganyika, GCP/RAF/271/FIN-FM/07. Bujumbura, Burundi: FAO.

PLISNIER, P. D. (1994) *Field Manual for the Second Year of Limnological Sampling on Lake Tanganyika*. FAO/FINNIDA Research for the Management of the Fisheries on Lake Tanganyika, GCP/RAF/271/FIN-FM/13. Bujumbura, Burundi: FAO.

PODSETCHINE, V. and HUTTULA, T. (1995) *Hydrological Modelling*. FAO/FINNIDA Research for the Management of the Fisheries on Lake Tanganyika, GCP/RAF/271/FIN-TD/29. Bujumbura, Burundi: FAO.

REYNOLDS, J. E. and GREBOVAL, D. F. (1988) *Socio-economic Effects of the Evolution of Nile Perch Fisheries in Lake Victoria: A Review*. Rome: FAO.

RICHARDSON, J. L. (1968) Diatoms and lake typology in East and Central Africa. *Int. Rev. gesamt. Hydrobiol. Hydrogr.*, **53** : 299–338.

ROBSON, P. J. and BAILEY-WATTS, A. E. (1987) A computer-based system for recording and sorting phytoplankton count data. *Brit. Phycol. J.*, **22** : 261–267.

ROUSSELET, C. F. (1910) Report on the Rotifera. Zoological results of the third Tanganyika Expedition (1904–1905). *Proc. zool. Soc. Lond.*, 792–796.

RUDD, J. W. M. (1980) Methane oxidation in Lake Tanganyika (East Africa). *Limnol. Oceanogr.*, **25** : 958–963.

RUFLI, H. (1976) *Preliminary Analysis on Zooplankton Sampling in Lake Tanganyika in May 1976*. FAO Report FI:DP/URT/71/012/44. Rome: FAO.

RUFLI, H. (1978) *Bibliography of Fisheries and Limnology for Lake Tanganyika*. FAO/CIFA Occasional Paper No. 5. Rome: FAO.

RUFLI, H. and CHAPMAN, D. W. (1976) *Preliminary Analysis of Zooplankton Sampling and Estimates of Fish Abundance in Lake Tanganyika in October 1975.* FAO Report FI:DP/URT/71/012/31. Rome: FAO.

RUFLI, H. and CHAPMAN, D. W. (1978) *Seasonal Changes in Zooplankton Abundance and Composition off Kigoma, Lake Tanganyika.* FAO Report FI:DP/URT/71/012/37. Rome: FAO.

SHACKLETON, N. J. (1980) Deep drilling in African Lakes. *Nature*, **288** : 211–212.

SHIMIZU, A (1987) Algal benthos in the north-western parts of Lake Tanganyika. pp. 111–113. In: KAWANABE, H. and NAGOSHI, M. (eds) *Ecological and Limnological Study on Lake Tanganyika and its Adjacent Regions*, IV. Kyoto, Japan: Kyoto University.

SINDAYIGAYA, E., VAN CAUWENBERGH, R., ROBBERECHT, H. and DEELSTRA, H. (1994) Copper, zinc, manganese, iron, lead, cadmium, mercury and arsenic in fish from Lake Tanganyika, Burundi. *Sci. Total Env.*, **144** : 103–115.

SOKAL, R. R. and ROHLF, F. J. (1969) *Biometry: the Principles and Practice of Statistics in Biological Research.* San Francisco: Freeman.

SPIVACK, A. and EDMOND, J. M. (1980) The ion balance of Lake Tanganyika. *Trans. Am. Geophys. Un.*, **61** : 46.

STOFFERS, P. and BOTZ, R. (1994) Formation of hydrothermal carbonate in Lake Tanganyika, east-central Africa. *Chem. Geol.*, **115** : 117–122.

SYMOENS, J. J. (1955a) Observation d'une fleur d'eau à Cyanophycées au Lac Tanganika. *Fol. sci. Afr. cent.*, **1** : 17.

SYMOENS, J. J. (1955b). Sur le maximum planctonique observé en fin de saison sèche dans le bassin nord du Lac Tanganika. *Fol. sci. Afr. cent.*, **1** : 12.

SYMOENS, J. J. (1956). Sur la formation de "fleurs d'eau" de la Cyanophycées (*Anabaena flos–aquae*) dans le bassin nord du Lac Tanganika. *Bulletin séanc. Acad. r. Sci. Colon., Bruxelles*, **2** : 414–429.

SYMOENS, J. J. (1959) Le développement massif de Cyanophycées planctoniques dans le Lac Tanganika. Abstract. In: *Proceedings of the 9th International Botanical Congress, Montreal*, 2A, No. 37.

TAKAMURA, K. (1987) Primary production of algae attached on rocks at Mbembe and Uvira coasts of Lake Tanganyika. p. 110. In: KAWANABE, H. and NAGOSHI, M (eds) *Ecological and Limnological Study on Lake Tanganyika and its Adjacent Regions*, IV.

TALLING, J. F. (1965) The photosynthetic activity of phytoplankton in East African lakes. *Int. Rev. Hydrobiol. Hydrogr.*, **50** : 1–32.

TALLING, J. F. (1986) The seasonality of phytoplankton in African lakes. *Hydrobiol.*, **138** : 139–160.

THORSLUND, A. E. (1971) *Report on Survey of Inland Water Pollution in Uganda, Kenya, Zambia and Tanzania.* FAO Regular Programme Report No. 11. Rome: FAO.

TIERCELIN, J.-J. (1988) Hydrothermal activity, metalliferous sediments and hydrocarbons. Examples of North Tanganyika and Kivu Troughs, East African Rift. Abstract presented at *International Workshop-Field Seminar, Lacustrine Facies Models in Rift Systems and Related Natural Resources*, Barcelona, October 1988.

TIERCELIN, J.-J., BOULEGUE, J. and SIMONEIT, B. R. T. (1993b) Hydrocarbons, sulphides, and carbonate deposits related to sublacustrine hydrothermal seeps in the North Tanganyika Trough, East

African Rift, pp. 96–113. *Special Publication* No. 9. Society for Geology Applied to Mineral Deposits.

TIERCELIN, J.-J., PFLUMIO, C., CASTREC, M., BOULEGUE, J., GENTE, P., ROLET, J., COUSSEMENT, C., STETTER, K. O., HUBERT, R., BUKU, S. and MIFUNDU, W. (1993a) Hydrothermal vents in Lake Tanganyika, East Africa Rift system. *Geology*, **21** : 499–502.

TIERCELIN, J.-J., THOUIN, C., KALALA, T. and MONDEGUER, A. (1989) Discovery of sublacustrine hydrothermal activity and associated massive sulfides and hydrocarbons in the north Tanganyika trough, East African Rift. *Geology*, **17** : 1053–1056.

TIERCELIN, J.-J., VASLET, N., THOUIN, C., KALALA, T., CHARLOU, J. L., FOUQUET, Y. and MONDEGUER, A. (1988) Découverte d'une activité hydrothermale à sulfures massifs en contexte d'ouverture intracontinentale. Les sites de Pemba et du Cap Banza, fosse nord-Tanganika, rift est-africain. pp. 138–140. In: *Colloque National sur l'Hydrothermalisme Océanique, Brest*, novembre 1988.

UNEP (1994) *The Pollution of Lakes and Reservoirs.* Nairobi, Kenya: United Nations Environment Programme.

VANDELANNOOTE, A., ROBBERECHT, H., DEELSTRA, H., VYUMVUHORE, F., BITETERA, L. and OLLEVIER, F. (in press) The impact of the River Ntahangwa, the most polluted Burundian affluent of Lake Tanganyika, on the water quality of the lake. *Hydrobiol.*

VANDELANNOOTE, A., VYUMVUHORE, F. and BITETERA, L. (1994) L'influence des rivières Ntahangwa et Mugere sur le Lac Tanganyika: les aspects physico-chémiques de la pollution macro-organique et de la pollution physique. p. 4. In: *Journées Scientifiques du Centre Régional de Recherches en Hydrobiologie Appliquée (CRRHA)*, 30–31 March 1994.

VASLET, N., TIERCELIN, J.-J., MONDEGUER, A. and THOUIN, C. (1987) Activité hydrothermale, sédiments metallifères et hydro-carbures. Le cas des fosses nord-Tanganika et Kivu rift est-africain. pp. 323–324. In: *Le Congrès Français de Sédimentologie, Paris*, novembre 1987.

VELGA, M. M. and MEECH, J. A. (1995) Gold mining activities in the Amazon: clean-up techniques and remedial procedures for mercury pollution. *Ambio*, **24** : 371–375.

VUORINEN, I. (1993) *Sampling and Counting Zooplankton on Lake Tanganyika.* FAO/FINNIDA Research for the Management of the Fisheries on Lake Tanganyika, GCP/RAF/271/FIN-FM/06. Bujumbura, Burundi: FAO.

VUORINEN, I. and KURKI, H. (1994) *Zooplankton Sampling on Lake Tanganyika.* FAO/FINNIDA Research for the Management of the Fisheries on Lake Tanganyika, GCP/RAF/271/FIN-TD/22. Bujumbura, Burundi: FAO.

WELLS, T. M., and COHEN, A. S. (1994) Human impacts and background controls of faunal diversity and community structure in Lake Tanganyika, Africa. *Geol. Soc. Am.*, **26** : 488.

WRIGHT, J. F., BLACKBURN, J. H., WESTLAKE, D. F., FURSE, M. T. and ARMITAGE, P. D. (1992) Anticipating the consequences of river management for the conservation of macroinvertebrates. pp. 137–149. In: BOON, P. J., CALOW, P. and PETTS, G. E. (eds) *River Conservation and Management.* Chichester: Wiley.

WRIGHT, J. F., FURSE, M. T. and ARMITAGE, P. D. (1993) RIVPACS – a technique for evaluating the biological quality of rivers in the UK. *Eur. Water Poll. Cont.*, **3** : 15–25.